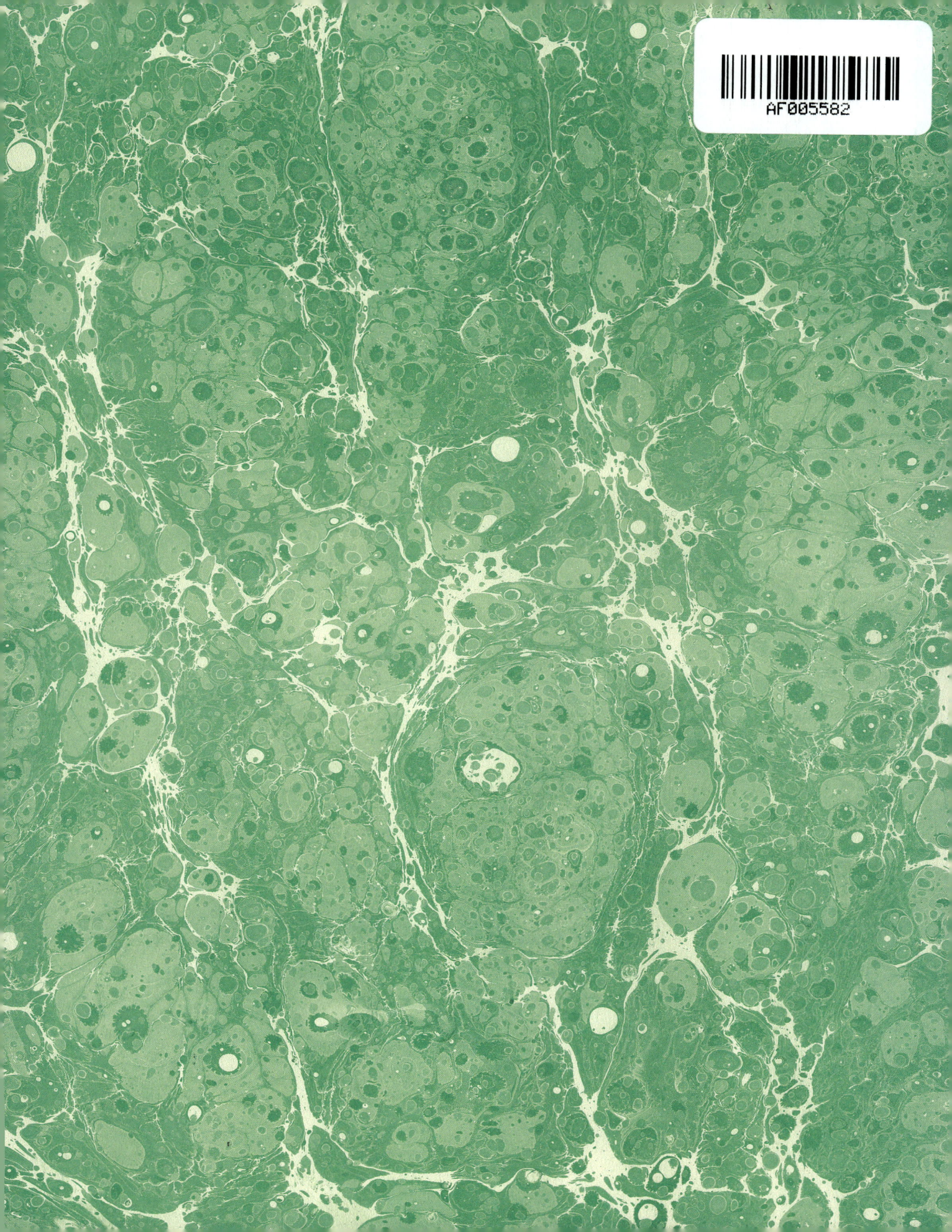

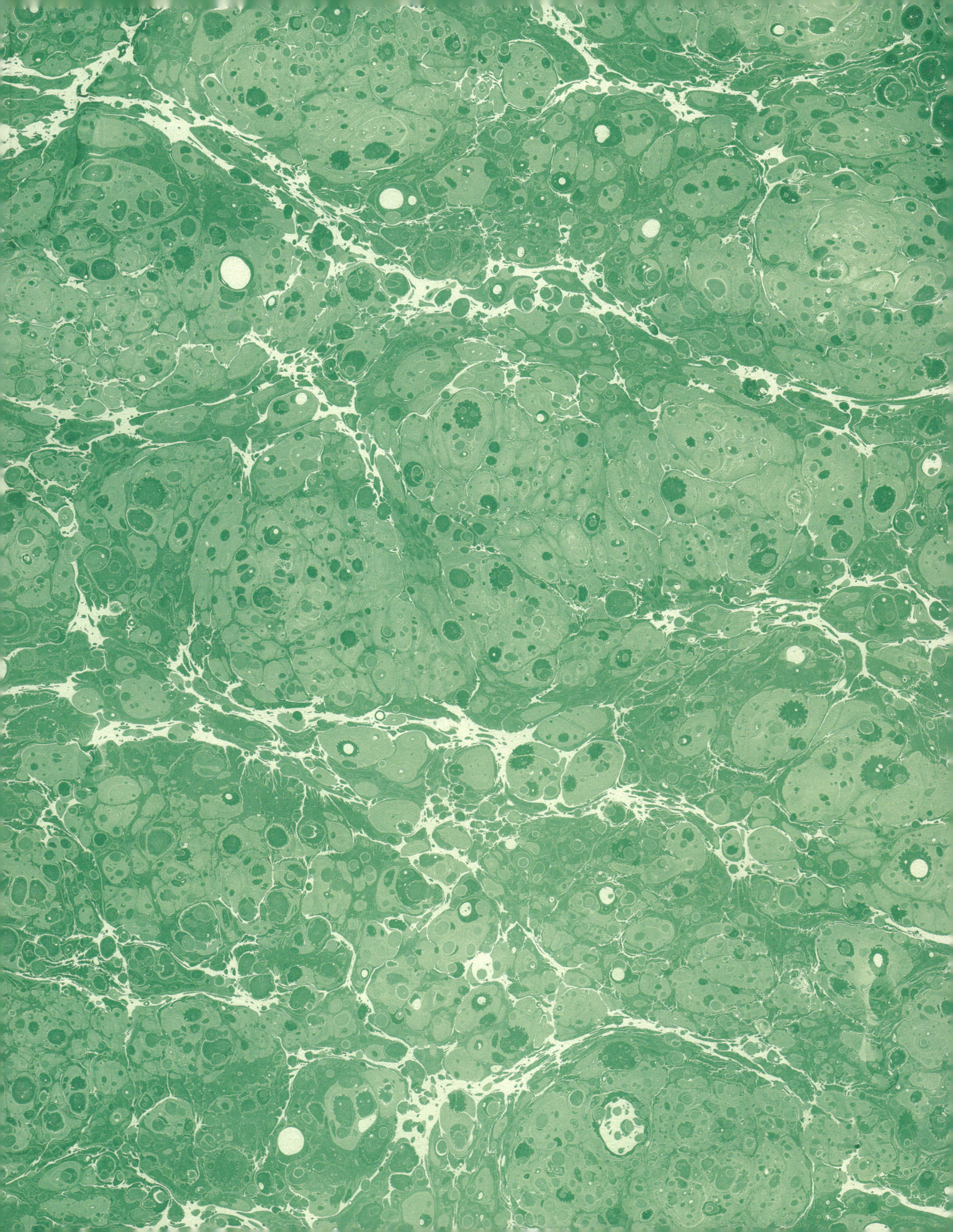

Aina Bestard

EATOPEDIA

An encyclopedia of how animals
eat, digest and poo

CONTENTS

I

What is the digestive system?   3

II

Digestive systems: 4
   Invertebrates   5
      Incomplete   6
      Complete   7
   Vertebrates   9
      Monogastric   10
      Birds   11
      Ruminants and pseudo-ruminants   12

III

Types of diet: 14
   Carnivores   15
   Herbivores   16
   Omnivores   16
   Other diets   17

IV

Seventy animals and their digestive systems   18

Glossary   170

I

What is the digestive system?

The majestic flight of an eagle, the buzzing of a mosquito on a summer's night, crabs dancing to attract mates, a cat miaowing to demand attention from its owner, a ferocious lioness chasing an antelope, and even what happens inside your brain as you read these few lines... All of these things and every other action that we animals perform, even the simplest ones we need to stay alive, like breathing or the beating of our hearts, require energy.

All animals, including humans, get energy from the food we eat. However, apart from some exceptional cases, there is a problem. The cells of our bodies can't get what they need directly from the food. We need a way to break down the nutrients in the food (protein, fat, carbohydrates, minerals, water and vitamins) and transform them into tiny particles that are small enough for the cells to use. The 'machine' that does this vital job is the digestive system.

The digestive process usually involves several stages that are common to nearly all animals: ingestion (which means the introduction of food into the body), digestion (which breaks the food down into smaller particles), absorption (when the broken-down nutrients are transported to the cells) and defecation (when the remains that have not been absorbed are expelled).

However, the animal kingdom is an incredibly varied world, with many different types of digestive systems that carry out these tasks in very different ways, depending on the animal. Also, the stages of the digestion process may be different in each animal, depending on the specific needs of the species. For some, the biggest challenge may be how to get the food to the digestive system, while for others, it's how to break down (or how to absorb) the nutrients from foods that are especially tough to digest. Even getting rid of waste raises challenges that are sometimes solved in very unusual ways. This is where our journey through some of the most fascinating digestive systems in the animal kingdom begins.

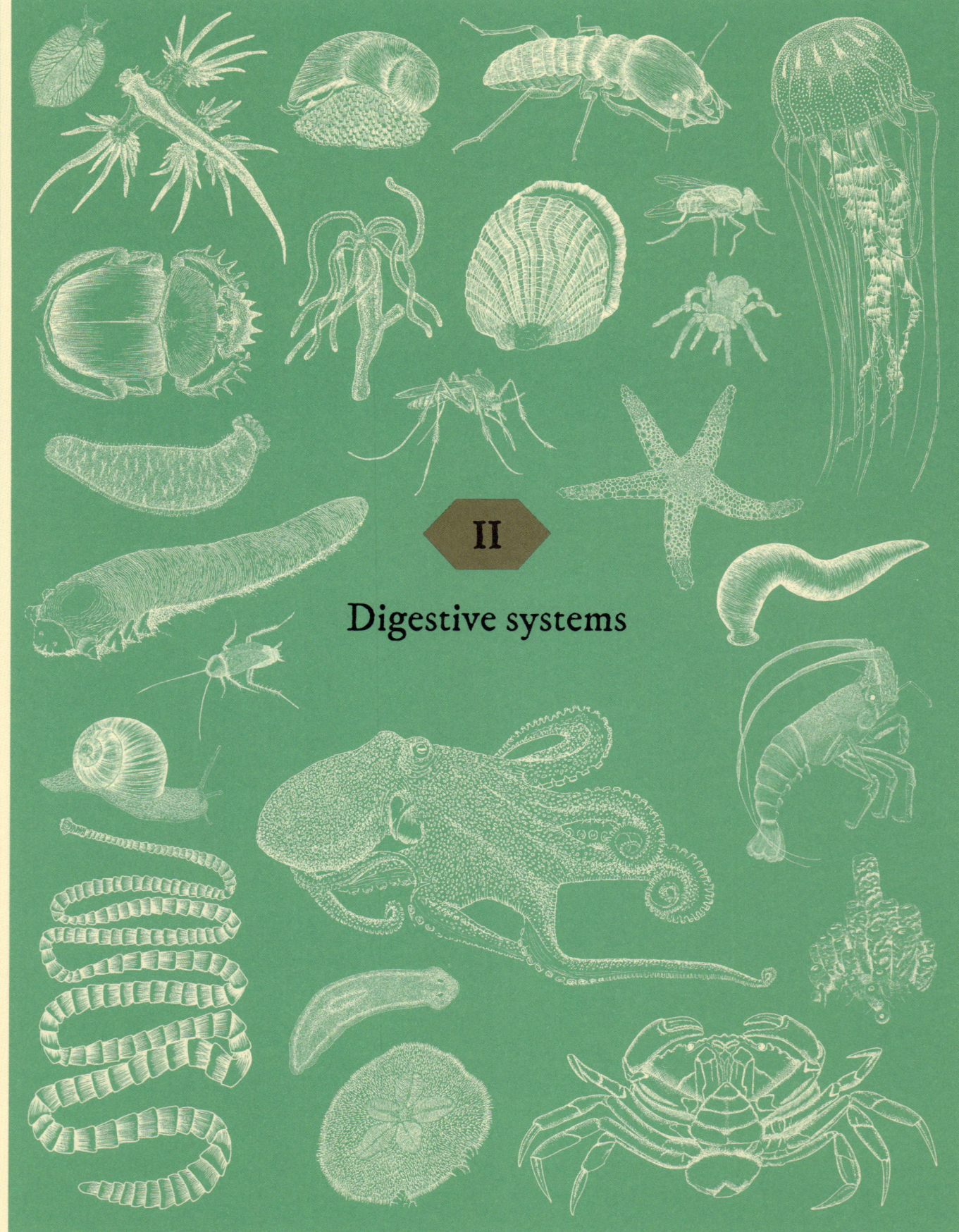

II

Digestive systems

INVERTEBRATES

Members of the animal kingdom can be divided into two main groups: vertebrates and invertebrates. Vertebrates have an internal skeleton that includes a spine, also known as a backbone. Invertebrates, on the other hand, make up a large majority of the Earth's animal species (well over 90% of them) and do not have a spine.

However, not having this type of internal skeleton is about the only thing that invertebrates have in common, since this group includes lifeforms that are very different from one another. For example, although invertebrates tend to be small, sometimes even microscopic (like many nematodes, a type of worm), there are also huge ones such as the giant squid, which can be up to ten metres long. Because they don't have bones, many invertebrates have soft bodies (again, think of worms). But some of them protect the soft parts of their body, either with a shell (like most molluscs) or with an exoskeleton, which is an outside skeleton (like insects, spiders and crustaceans). There are even invertebrates like the starfish, which have a different kind of internal skeleton, not made of bone but from other materials.

The group of invertebrates includes the simplest forms of life in the animal kingdom although there are major differences between them. The more primitive ones don't have organs or even entire organ systems – for example, sea sponges don't have a brain or blood – that we often take for granted when we think of animals. You won't be surprised then that the digestive systems of invertebrates are generally simpler than those of vertebrates. Let's take a closer look!

II.
Digestive systems

INVERTEBRATES

INCOMPLETE

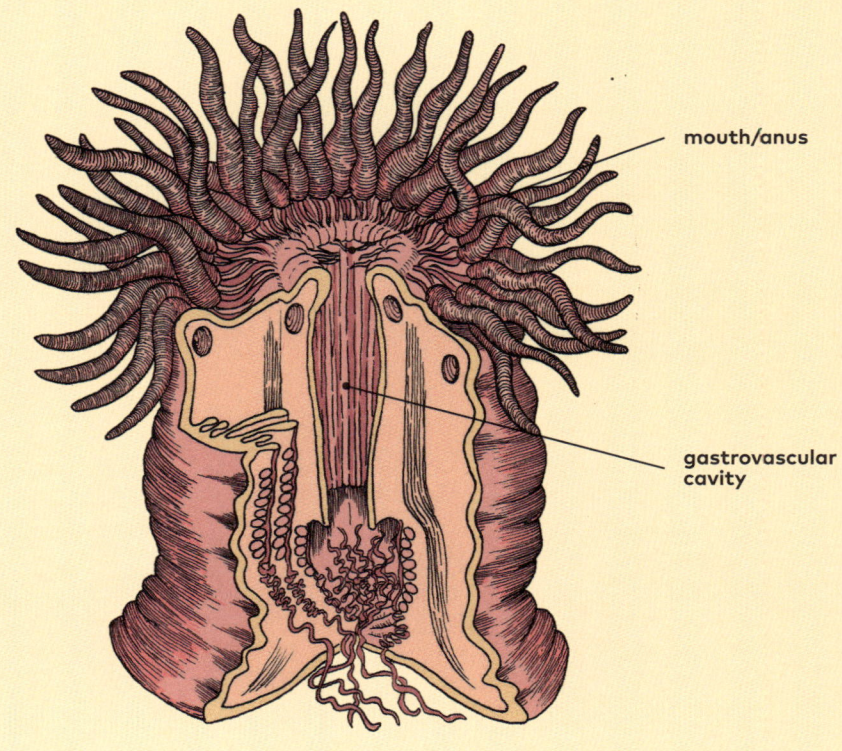

BEADLET ANEMONE
Actinia equina

Some invertebrates don't have all the parts of a typical digestive system. They have developed other ways of getting what they need, without some of the organs that carry out the usual jobs of the digestive tract: ingestion (eating), digestion, absorption, and excretion (getting rid of waste).

In a few cases, like sea sponges, digestion is done by the individual cells that make up the body. This is called intercellular digestion. Instead of eating food, processing it, and breaking down its nutrients, which are then taken to the cells, these animals take in particles that contain the substances they need, and the cells themselves do the job of digesting them. It couldn't be any simpler! Some invertebrates that are parasites may not have a digestive system because they get all the nutrients they need directly from the host creature that they live on.

Most invertebrates with an incomplete digestive system are slightly more complex than that, however. Cnidarians (a group that includes jellyfish, corals and sea anemones) and platyhelminths (also known as flatworms) have a basic organ called a gastrovascular cavity. It is a type of bag that can digest and absorb the food they take in. This organ has only one entrance: it's where food comes in and where waste is pushed out. In other words, it works as both mouth and anus at the same time. This sounds very strange to us but it works for them!

II.
Digestive systems

INVERTEBRATES

COMPLETE

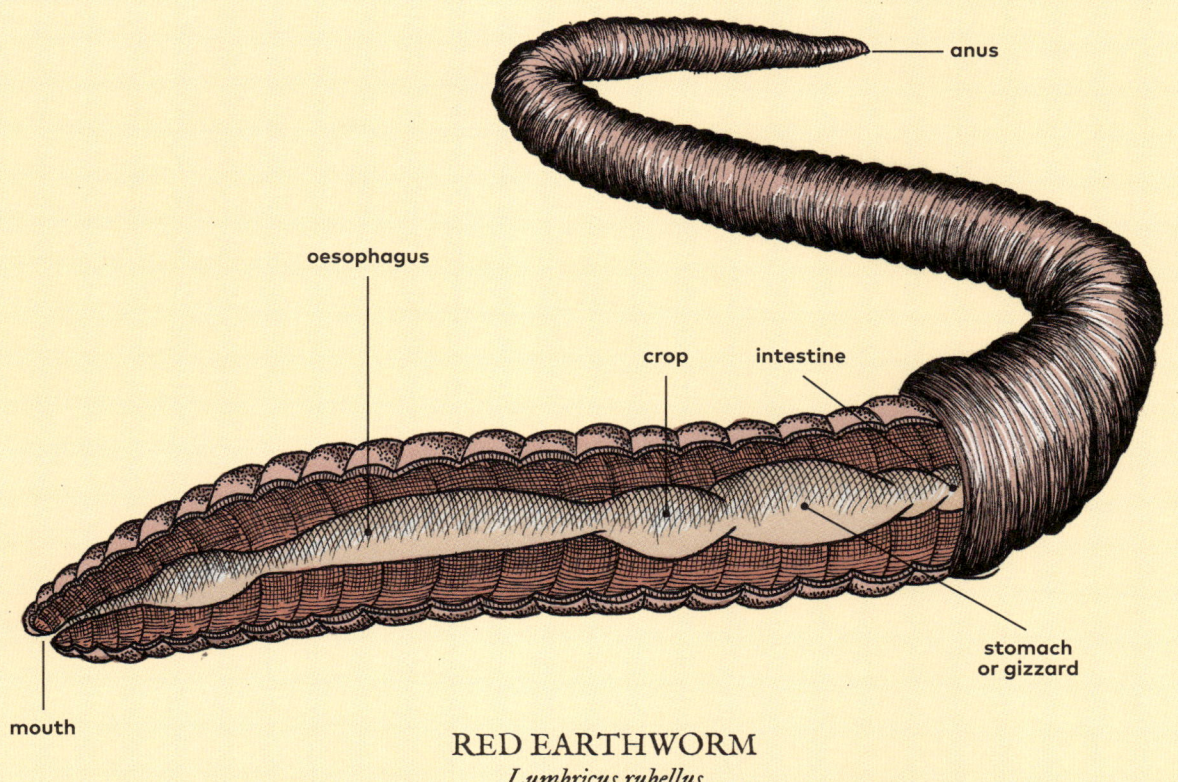

RED EARTHWORM
Lumbricus rubellus

Most invertebrates have a complete digestive system, which is similar to the ones that humans and other vertebrates have. It consists mainly of a long tube, known as the alimentary canal or digestive tract. This tube has an entrance at one end (the mouth) and an exit at the other (the anus).

In this type of system, the digestive tract is divided into sections. Each one has a different job to do. For example, one section may store food (the crop), one grinds it up (the stomach or gizzard), one breaks it down into nutrients and absorbs them (the intestine), and one stores waste until it is time to be expelled from the body. The alimentary canal may sometimes include glands (the salivary glands, or the hepatopancreas glands, which are like our pancreas and our liver). These produce substances that help to break down the food into chemicals.

These specialist sections mean that having a complete digestive system is more efficient. With its help, invertebrates such as arthropods (insects, spiders, crustaceans), annelids (earthworms and many other worms) and molluscs (clams, octopuses, snails) have evolved into a huge number of species that can adapt to a very broad range of environments and eat a much wider range of foods than invertebrates with incomplete digestive systems.

VERTEBRATES

Vertebrates have an internal skeleton that includes a spine, also called a backbone. This is the group that we humans belong to, along with all other mammals, as well as birds, fish, reptiles and amphibians. Vertebrates are more complex than invertebrates, and this is reflected in their digestive systems.

The usual digestive system of a vertebrate begins with a mouth. This generally includes a tongue, teeth or a beak, and in land animals, saliva glands. The mouth is used to ingest (take in) food. Vertebrates often grind up their food by chewing it and start to break it down with their saliva. The food is then carried down the pharynx (throat) and oesophagus to the stomach.

The stomach is a stretchy bag of muscle that mixes and squeezes food. It is filled with acids (the gastric juices), which break down the food into smaller particles.

In the intestine, the food is broken down even more, its nutrients are absorbed, and waste material is left behind. In many vertebrates, the intestine is divided into two sections. The first is the small intestine, which finishes breaking down the food into nutrients and absorbs the most simple molecules. This job is done with the help of substances made by other organs such as the liver and pancreas. The second section is the large intestine. This is where water and more complex nutrients are absorbed, with the help of the microorganisms living there. It is also where the leftover waste (faeces or poo) is gathered, before it is pushed out through the anus or cloaca.

Although this is the basic pattern, there can be big differences, depending on the needs and diet of each species. Vertebrate digestive systems can be divided into four types, so let's look at them in more detail!

II.
Digestive systems

VERTEBRATES

MONOGASTRIC

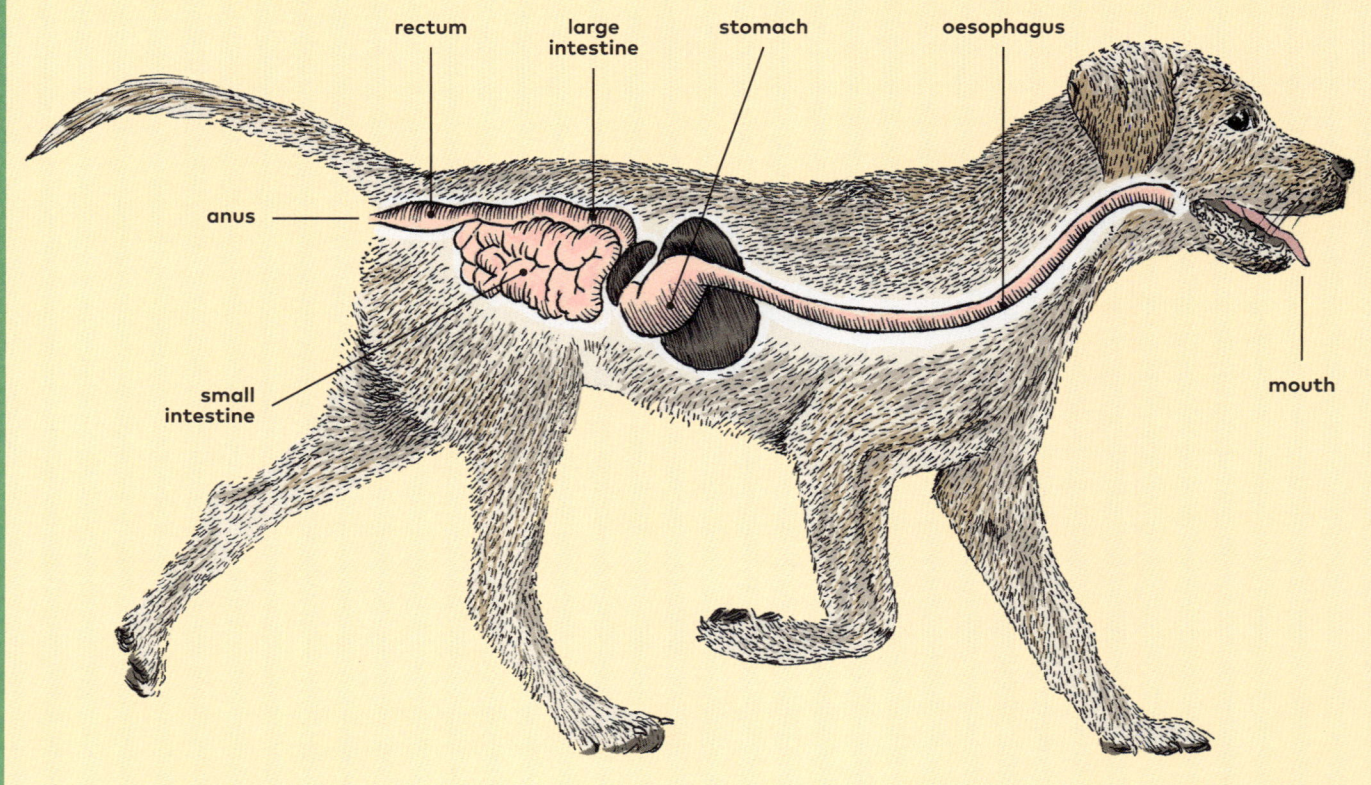

DOG
Canis lupus familiaris

The word monogastric means 'one stomach'. This type of digestive system has a simple stomach with just one chamber.

Human beings and many other mammals (dogs, cats, rabbits, pigs) have monogastric digestive systems. Fish, reptiles and amphibians also have digestive systems with a single stomach chamber, although with some differences that reflect their needs.

Monogastric systems are good for animals that eat high-energy foods which are relatively easy to digest, like meat, fruit, and the most nutritious parts of plants.

In contrast, getting nutrients from very high-fibre plants is more complicated with this type of digestive system. However, some animals, such as rabbits, manage to do it.

II.
Digestive systems

VERTEBRATES

BIRD

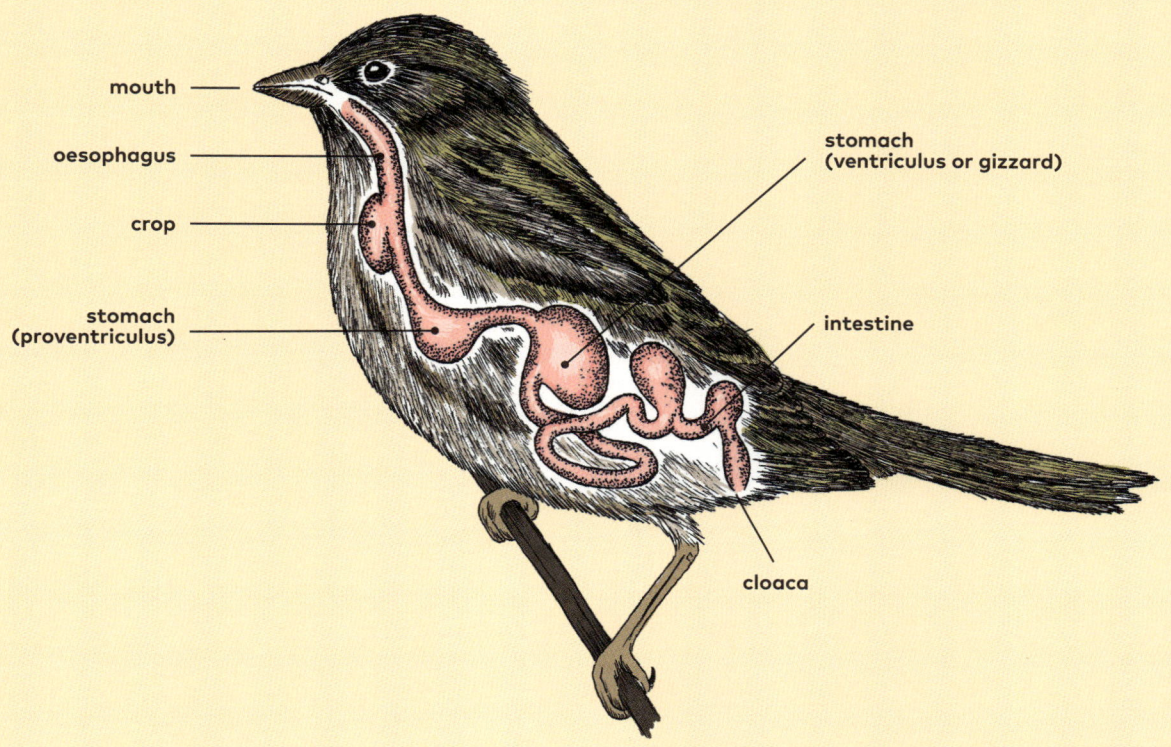

COMMON CANARY
Serinus canaria

Birds are the only animals with feathers, which many of them use to fly. This ability has allowed birds to adapt to all kinds of habitats and find many different sources of food, but it also means that they need a lot of energy. They must eat often and digest quickly. Their mouths have beaks but no teeth, which means they can't chew food. All they can do is break it up with their beaks, so they often swallow food whole or almost whole. Their digestive system has adapted to deal with this.

In order to always have energy reserves, many birds have a chamber called a crop (seen also in some insects) that acts as a storage area for food so it can be digested later. A bird's stomach is divided into two parts: the proventriculus and the ventriculus or gizzard. The first produces gastric juices, while the second grinds up the food, often with the help of sand or stones the bird has swallowed. This is how they make up for their missing teeth: 'chewing' using stones in their stomach!

Finally, after the food passes through the intestine, the waste is pushed out through the bird's cloaca. A cloaca is similar to an anus: the difference is that both faeces (poo) and urine (pee) are pushed out through the same opening. This explains why bird droppings are usually very liquid!

II.
Digestive systems

VERTEBRATES

RUMINANTS AND PSEUDO-RUMINANTS

Some parts of plants, such as the roots and leaves, contain very few nutrients and are hard to digest because they contain a lot of plant fibre called cellulose (this substance is also used to make paper). However, pseudo-ruminants and ruminants are mammals that have evolved to be very good at digesting this kind of food.

Ruminants include cows, goats, sheep and deer, among others. They have a stomach that is divided into four chambers: these are called the rumen, reticulum, omasum and abomasum. They eat large amounts of food without chewing it and store it in the rumen, the first part of their stomachs. The rumen is the largest of the four chambers and contains thousands of millions of microorganisms that do what gastric juice cannot do: break down cellulose. These microorganisms, which are mainly bacteria and protozoa, make the food ferment and the ruminant animal is then able to absorb some of its nutrients.

Later, the animal regurgitates the food – this means that it brings it back up into its mouth. Then it chews the food again, taking its time. It only does this when it feels safe and there are no predators around. This is called 'rumination' or 'chewing the cud'.

After rumination, the food is swallowed again and passes through the rumen to the other chambers of the animal's stomach. The last of the four chambers is the abomasum, which is the one most like a human stomach. This is where gastric juices do their job. Then, just like in the digestive systems of other vertebrates, the food passes through the small and large intestines, and the waste comes out through the anus.

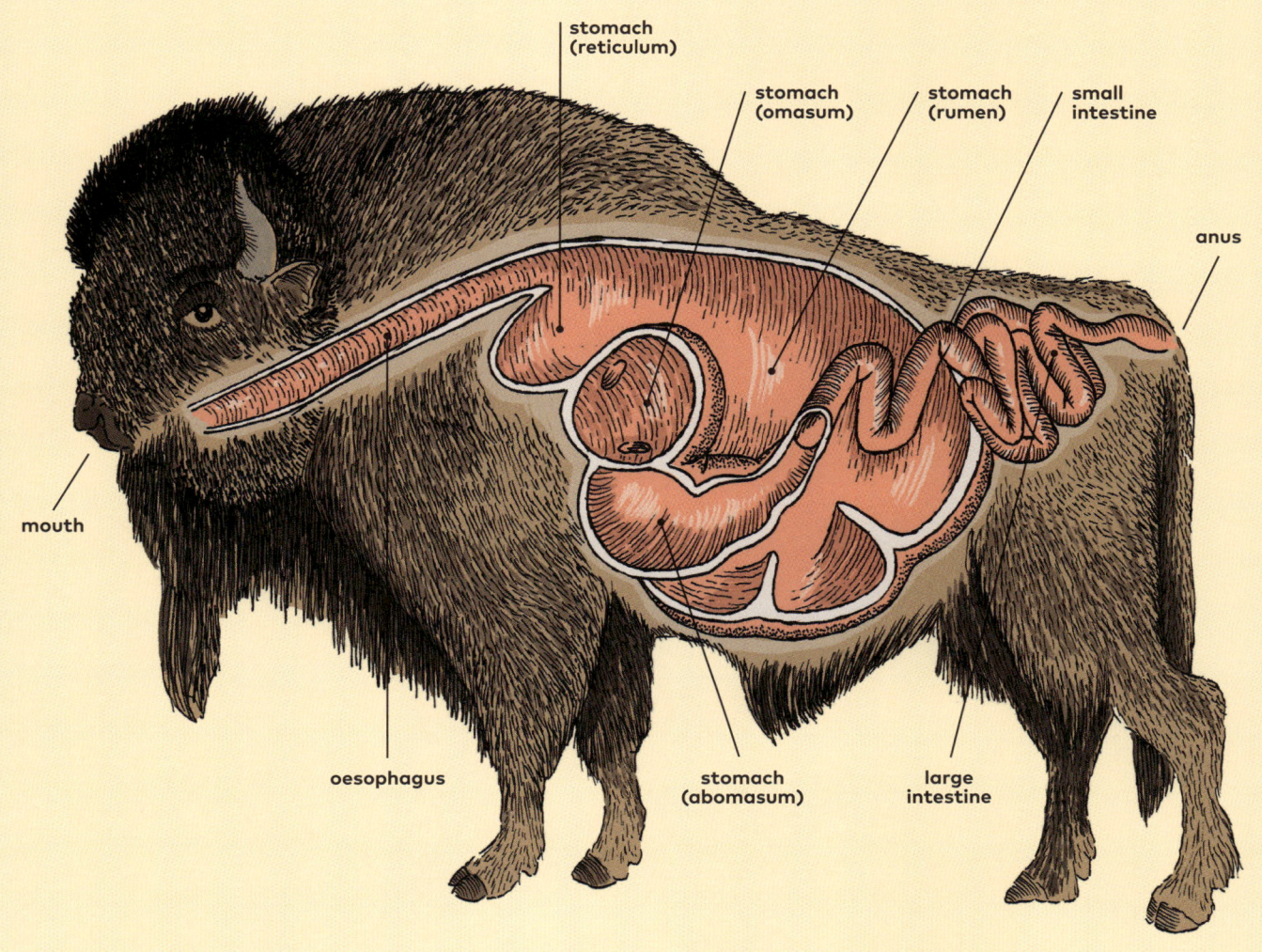

AMERICAN BISON
Bison bison

Pseudo-ruminants

The final group of vertebrates contains animals such as camelids (a family that includes species like the camel, alpaca and llama) and hippopotamuses. These have a digestive system similar to ruminants, but their stomach has three chambers instead of four. They are known as pseudo-ruminants.

III

Types of diet

III.
Types of diet

The way an animal's digestive system works is heavily influenced by their diet. The organs of each species have evolved to swallow, digest, absorb and excrete the foods that each particular animal prefers.

Animals can be divided into three main groups, according to the food they eat: carnivores, herbivores and omnivores.

CARNIVORES

Carnivores are animals who eat other animals. Another term for them is zoophagous animals. Perhaps confusingly, there is a group of mammals called the order Carnivora that mainly feed on meat: this group includes wolves, dogs, cats, lions, tigers and more. Nearly all members of the order Carnivora are carnivores, but there are lots of carnivores that are not part of the order Carnivora.

Most carnivores are predators: they hunt their own prey themselves. However, some animals are necrophages, which means that they eat carrion: the remains of dead animals killed by other creatures. But both of these groups eat other animals, which means that their digestive systems are pretty similar.

Firstly, their mouths have evolved to suit the type of animal they eat. For example, animals that eat meat need sharp teeth to rip it up, while birds of prey have hooked beaks to help them tear up flesh. Alternatively, animals that eat insects have mouths that have adapted to catch them easily: examples include the anteater's long tongue or birds with sharp tweezer-like beaks.

As their food is relatively easy to digest, carnivores have a digestive tract that is relatively short: around three to six times as long as their body. The largest organ in it is the stomach, which makes up 60% to 70% of the system's capacity. Animals with an insect-based diet have even simpler digestive tracts.

III.
Types of diet

HERBIVORES

Herbivores, also called phytophagous animals, eat plants and all their parts: roots, seeds, stems, flowers, nectar, sap, leaves, fruits and even wood.

Herbivorous mammals include the ruminants and their cousins, the pseudo-ruminants. We have already seen that these animals have very complex digestive systems that can digest the cellulose of plants they eat. There are also herbivores with simpler (monogastric) digestive systems, but they often have a long section in their large intestine called the caecum. This helps them to digest cellulose.

Mammals who eat cellulose-rich plants tend to have the longest digestive tracts in nature, sometimes up to ten times the length of their bodies. In contrast, mammals that eat other plant parts, such as fruit, have shorter digestive tractss because those parts are easier to digest.

Insects are another major group of herbivores. There are lots of species in this group and some of them specialize in eating specific parts of plants, such as termites that feed on wood, or bees that feed on pollen and nectar.

III.
Types of diet

OMNIVORES

Omnivores are animals that eat both plants and other animals. However, it is true that many species eat both plants and animals occasionally: herbivores accidentally eat insects on the plants they eat, and many carnivores also eat plants from time to time, as cats do with grass.

So how do we sort animals into categories? Well, when their diet is heavily based on one of the two types of food, we classify them as carnivores or herbivores. Animals with a combined diet that don't fit neatly into either of the previous two groups are known as omnivores. This is the category that human beings belong to! Pigs, crows and most species of bears are also omnivores.

The digestive systems of omnivores have characteristics that are somewhere between those of herbivores and carnivores, and they vary according to the species. The digestive system of bears, for example, looks a lot more like a carnivore's digestive system, while the human digestive system has more in common with herbivores.

III.
Types of diet

OTHER DIETS

There are also special terms for animals with diets based on a very specific type of food.

Haematophages are animals that feed on the blood of other living animals. You might say they are nature's original vampires! When you think of animals that drink blood, you may immediately think of bats. However, although there are more than one thousand species of bats in the world, only three species actually drink blood. Those three come from the subfamily Desmodontinae, and are better known as vampire bats. Blood-drinking is much more common among invertebrates, such as mosquitoes, fleas, ticks, lice and leeches. There are also haematophage fish (e.g. lampreys) and birds (e.g. Darwin's finches).

If you think feeding on blood sounds unpleasant, wait until you find out about our next group of animals: the coprophages. The term comes from the Greek words for 'faeces' and 'to eat', so perhaps you can already guess where this is going. Yes, these animals feed on poo! Coprophagia is pretty common in nature. There are many species that eat poo occasionally: elephants, rabbits, hares, dogs and even primates such as lemurs. However, the animals that feed most heavily on poo are insects: these include dung beetles and some species of flies.

There are many more feeding strategies that have their own names. Detritivores (including sea sponges and sea cucumbers) eat rotting organic matter (detritus). Osteophages (like the bearded vulture) feed on bones. Chemosynthetic species (like the volcano snail) have bodies that can turn surrounding molecules into energy, similar to the way that plants use photosynthesis to create energy from sunlight. Animals who feed only on ants and termites (like anteaters) are called myrmecophages, and those that eat only plankton (microscopic organisms that float in the water) are planktivores. There are even animals called lithophages that feed on rock, such as the shipworm (*Lithoredo abatanica*).

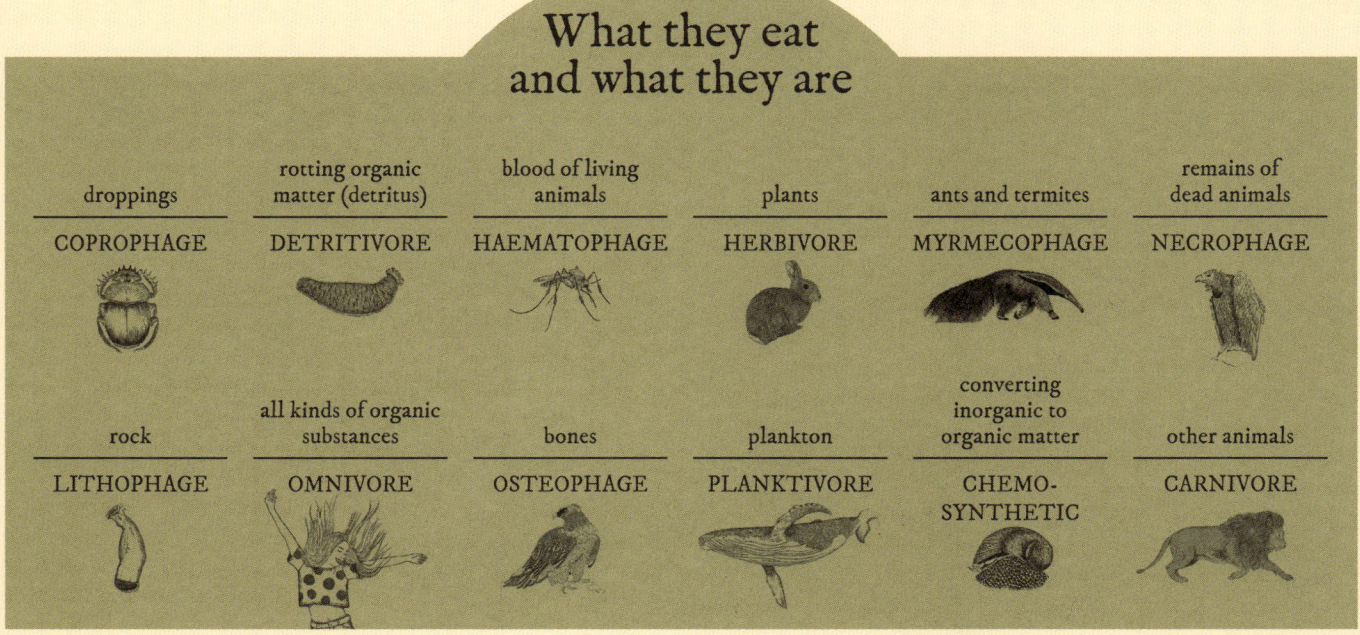

What they eat and what they are

droppings	rotting organic matter (detritus)	blood of living animals	plants	ants and termites	remains of dead animals
COPROPHAGE	DETRITIVORE	HAEMATOPHAGE	HERBIVORE	MYRMECOPHAGE	NECROPHAGE

rock	all kinds of organic substances	bones	plankton	converting inorganic to organic matter	other animals
LITHOPHAGE	OMNIVORE	OSTEOPHAGE	PLANKTIVORE	CHEMO-SYNTHETIC	CARNIVORE

INVERTEBRATES
Incomplete digestive system — 20
1. Crab hacker barnacle — 22
2. Planarian — 24
3. Sea sponge — 26
4. Brown hydra — 28
5. Tapeworm — 30
6. Sea nettle jellyfish — 32

INVERTEBRATES
Complete digestive system — 34
7. Starfish — 36
8. Green sea slug — 38
9. Common spiny lobster — 40
10. Tarantula — 42
11. Honey bee — 44
12. Medicinal leech — 46
13. Dung beetle — 48
14. Sea cucumber — 50
15. Blue sea dragon — 52
16. Volcano snail — 54
17. Mosquito — 56
18. Common octopus — 58
19. Shipworm — 60
20. Black cockroach — 62
21. Pacific sand dollar — 64
22. Housefly — 66
23. Termite — 68
24. Face mite — 70
25. Roman snail — 72
26. Black-lip pearl oyster — 74

VERTEBRATES
Birds — 76
27. Barn owl — 78
28. Bee hummingbird — 80
29. Domestic pigeon — 82
30. Greater flamingo — 84
31. Bearded vulture — 86
32. Bald eagle — 88
33. Hoatzin — 90
34. Ostrich — 92
35. Chicken — 94

VERTEBRATES
Monogastric — 96
36. Human — 98
37. Red kangaroo — 100
38. Domestic cat — 102
39. Asian palm civet — 104
40. Great white shark — 106
41. Common remora — 108
42. Platypus — 110
43. Wombat — 112
44. Dog — 114
45. Porcupine pufferfish — 116
46. Common anaconda — 118
47. Giant panda — 120
48. Horse — 122
49. European rabbit — 124
50. Giant anteater — 126
51. Common vampire bat — 128
52. Mountain tree-shrew — 130
53. Three-toed sloth — 132
54. Bluestreak cleaner wrasse — 134
55. Blue whale — 136
56. Hagfish — 138
57. Domestic pig — 140
58. Chimpanzee — 142
59. Lion — 144
60. Brown rat — 146

VERTEBRATES
Ruminants and pseudo-ruminants — 148
61. Java mouse-deer — 150
62. Cow — 152
63. Giraffe — 154
64. Domestic sheep — 156
65. Red deer — 158
66. Iberian ibex — 160

67. Hippopotamus — 162
68. Bactrian camel — 164
69. Llama — 166
70. Vicuña — 168

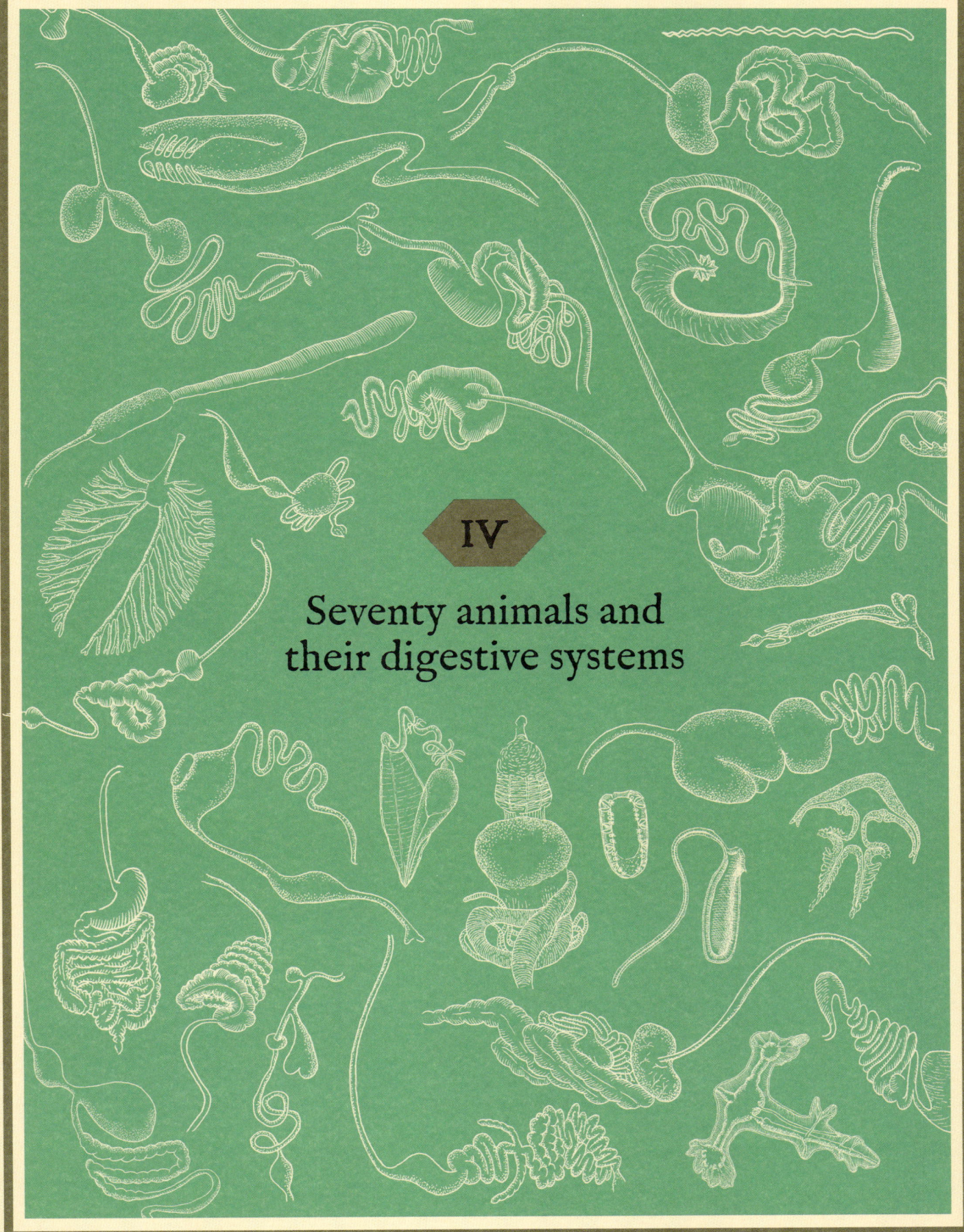

IV

Seventy animals and their digestive systems

1.
Crab hacker
barnacle

2.
Planarian

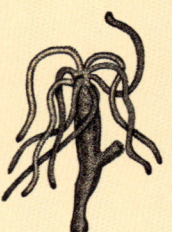

3.
Sea sponge

4.
Brown hydra

5.
Tapeworm

6.
Sea nettle jellyfish

INVERTEBRATES

Incomplete digestive system

This creature is a parasite, which means that it lives on another species, called a host. Its chosen host is the European green crab (*Carcinus maenas*).

1

Only the female barnacles grow into this parasitic adult form. Males stop growing at a larval stage.

CRAB HACKER BARNACLE
Sacculina carcini

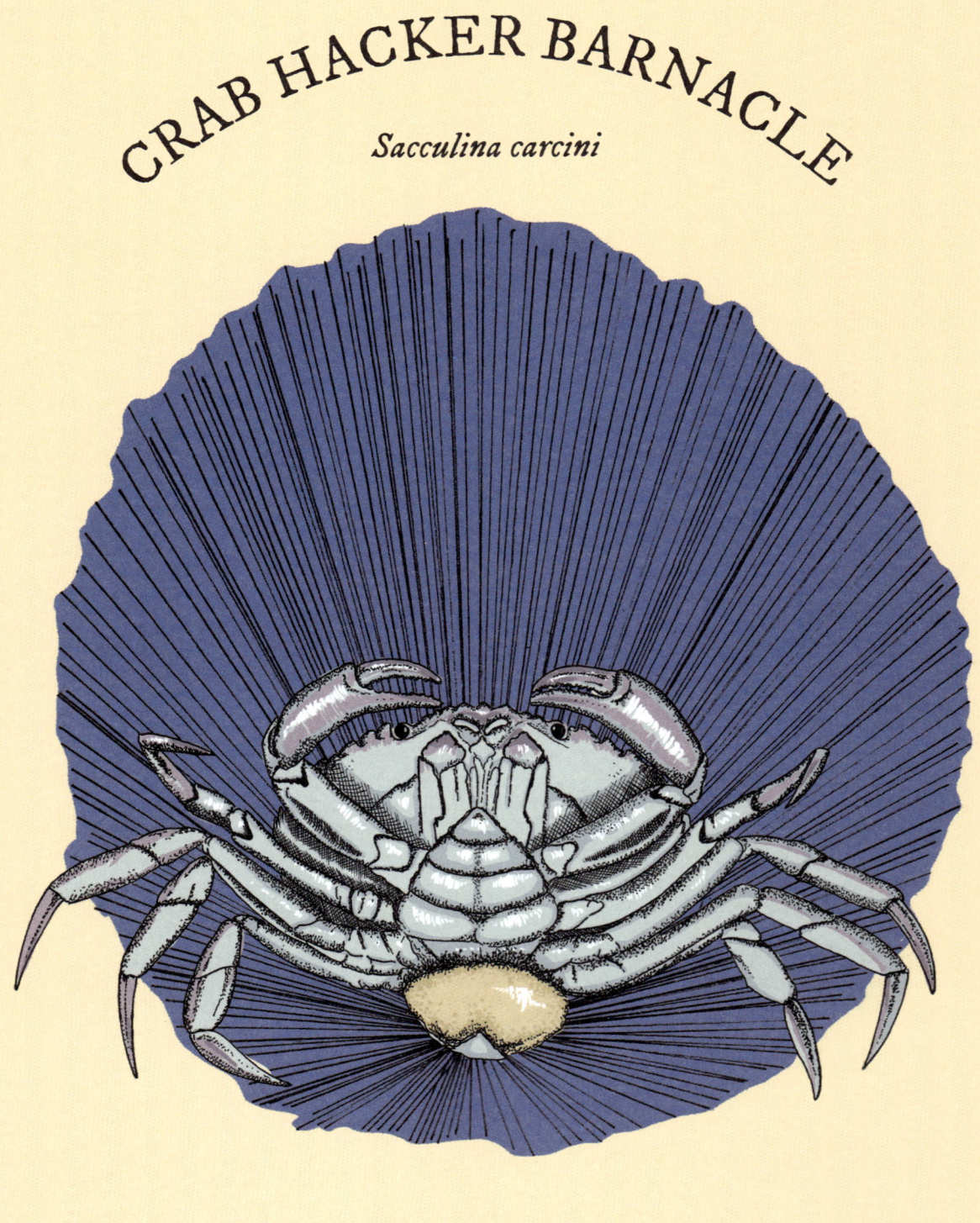

Animal:
INVERTEBRATE

Digestive system:
INCOMPLETE

Type of diet:
PARASITE

A born scrounger!

What is the easiest way to digest food? For the crab hacker barnacle, the answer is to avoid growing a digestive system of your own, and borrow another animal's instead! This parasite lives on a crab and feeds on the food that the crab has eaten and digested.

The barnacle absorbs the nutrients it needs directly from the host crab's haemolymph. This is the crustacean equivalent of blood.

The crab hacker barnacle does this by growing a system of root-like tendrils that spread throughout the host's body, focusing on its digestive and nervous system.

FAVOURITE FOOD
Anything eaten by the crab.

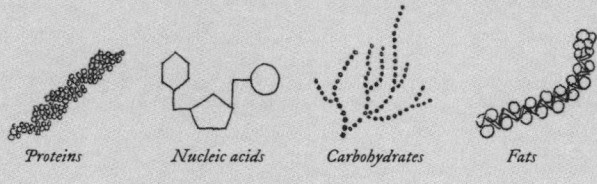

Proteins *Nucleic acids* *Carbohydrates* *Fats*

WHEN DO THEY EAT?
Whenever the crab eats. That's the good thing about being a parasite: your host does all the work for you.

WHO EATS THEM?
Anything that eats the host crab. This includes birds, fish, lobsters and other crabs.

1. root-like tendrils
2. body of the barnacle
3. barnacle's reproductive system

EAT TO REPRODUCE

The tendrils of the crab hacker barnacle take the nutrients to the part of the parasite that hangs outside the crab's body, underneath its shell. This part is mostly a reproductive system, so the barnacle is really just a machine for reproducing itself.

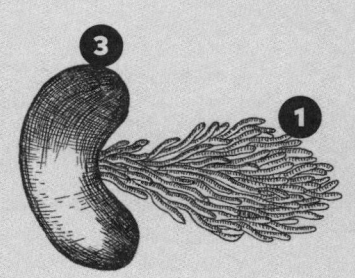

ZOMBIE CRABS!

The barnacle makes the host crabs infertile and controls their behaviour, as if they were zombies. For example, it makes the males think they are females! The host crab behaves as if the barnacle's reproductive system was its own and pushes out the barnacle larvae as if they were its own eggs.

Planarians are a type of flatworm, one of the oldest and simplest families of invertebrates.

2

They live in water, which most often means the sea, although some live in freshwater areas or in wet places on land.

PLANARIAN
Turbellaria

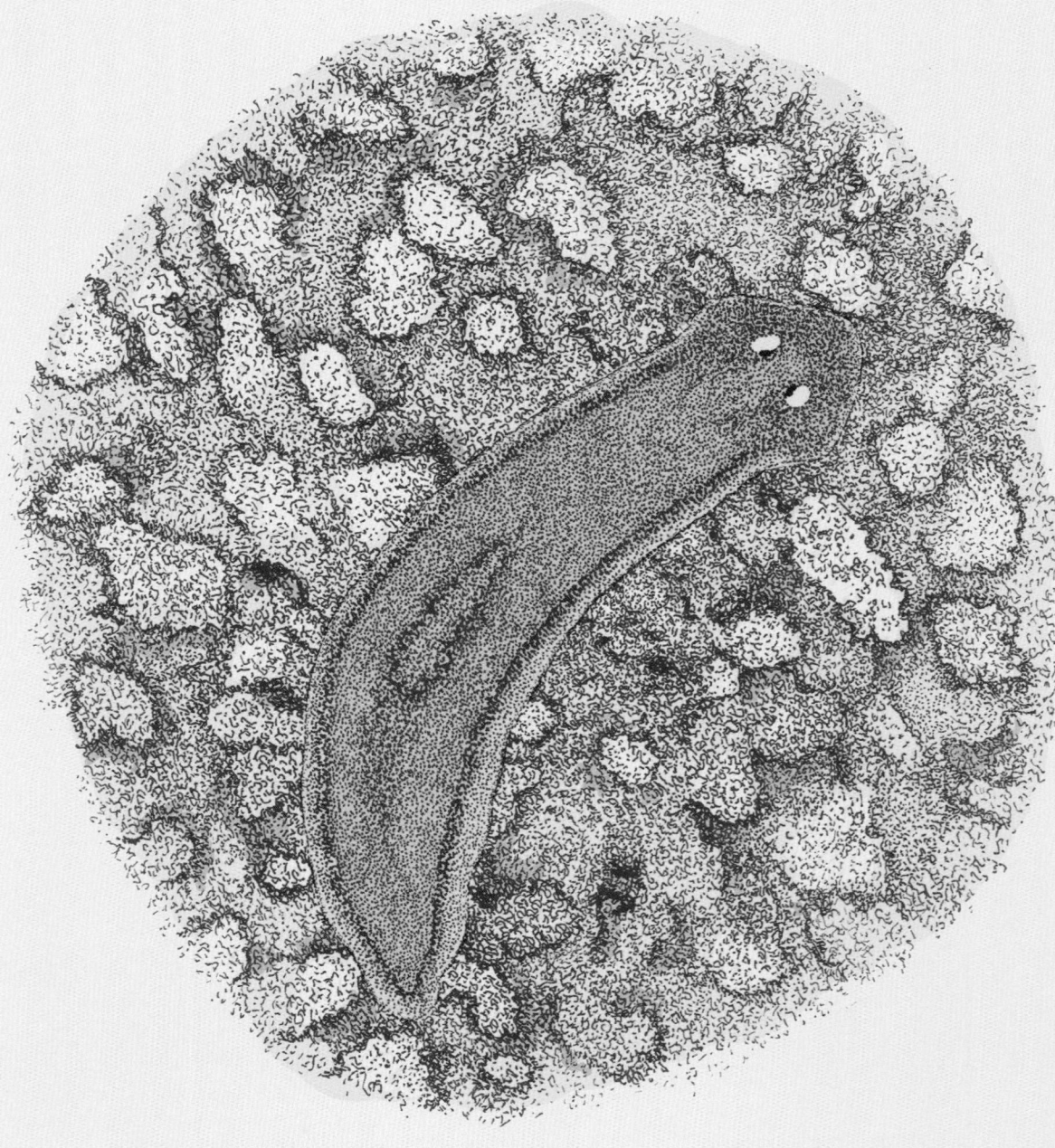

Animal:
INVERTEBRATE | Digestive system:
INCOMPLETE | Type of diet:
CARNIVORE

No way out!

A planarian is a good example of an incomplete digestive system. Its mouth is on the underside of its body, in the centre. From there, a powerful pharynx (throat) carries food to the worm's intestine. This branches through the planarian's body, but there's no anus to act as an exit. Any undigested food left over must be pushed out through the same place it went in: the mouth.

FAVOURITE FOOD

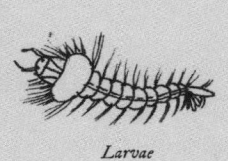
Larvae

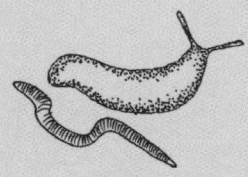
Worms and slugs smaller than they are

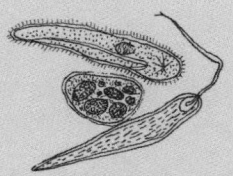
Protozoa

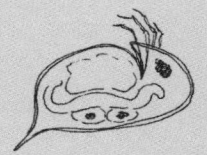
Small crustaceans and molluscs

Most planarians are carnivores that feed on other small water creatures.

Their skin cells produce a sticky mucus that paralyses their prey.

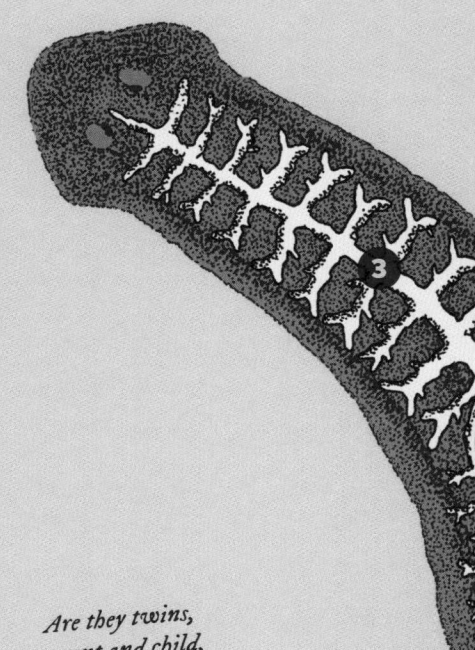

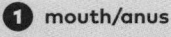

1. mouth/anus
2. pharynx (throat)
3. digestive tract

WHEN DO THEY EAT?

Planarians tend to hunt for prey at night. They may be flat, but they're fast movers!

WHO EATS THEM?

The mucus they make also has another purpose: it makes planarians less tasty to predators. But this doesn't stop fish, amphibians and even some insect larvae from eating them.

The last stage of digestion happens in the planarian's cells. Any waste is expelled through the planarian's skin.

Are they twins, parent and child, or clones?

ONE WORM BECOMES TWO

Perhaps the most remarkable thing about planarians is one of the ways they can reproduce. It's called transversal division. In this process, the worm splits its body in half and each one of the halves regenerates to form a complete worm. This can also happen if the worm gets cut in half by accident!

DELIVERY SERVICE

The digestive tract of a planarian forms lots of tiny branches inside its body. This is because these creatures are so simple that they don't have veins, arteries or blood to carry nutrients to the body's cells. So it's up to the digestive system itself to do the job as best it can.

Sponges are one of the oldest groups of animals in the world. They've been on the planet for more than 600 million years.

3

For a long time, people believed that sponges were plants because they look like plants and don't move. In fact, some of them do move, just very slowly.

SEA SPONGE
Porifera

Animal:
INVERTEBRATE

Digestive system:
INCOMPLETE

Type of diet:
DETRITIVORE

Straight to the cells

Sponges are a very primitive life form. They don't have the internal organs that most other animals have, such as a digestive system. Instead, their individual cells do the jobs that those organs would usually do. To feed themselves, the sponges' cells take in nutrients and oxygen directly from the water they live in. This is called intracellular digestion.

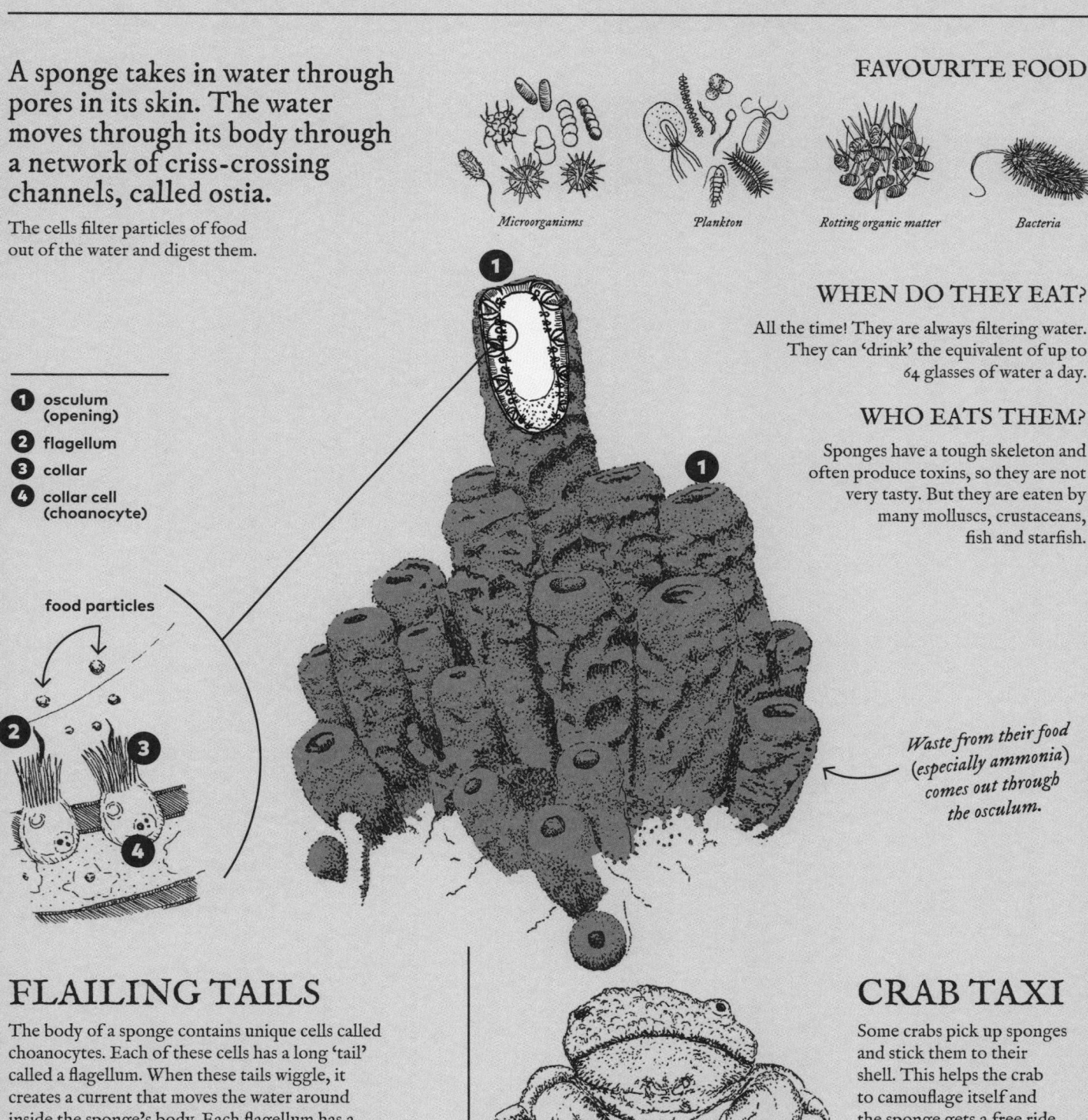

A sponge takes in water through pores in its skin. The water moves through its body through a network of criss-crossing channels, called ostia.

The cells filter particles of food out of the water and digest them.

1. osculum (opening)
2. flagellum
3. collar
4. collar cell (choanocyte)

food particles

FAVOURITE FOOD

Microorganisms *Plankton* *Rotting organic matter* *Bacteria*

WHEN DO THEY EAT?
All the time! They are always filtering water. They can 'drink' the equivalent of up to 64 glasses of water a day.

WHO EATS THEM?
Sponges have a tough skeleton and often produce toxins, so they are not very tasty. But they are eaten by many molluscs, crustaceans, fish and starfish.

Waste from their food (especially ammonia) comes out through the osculum.

FLAILING TAILS
The body of a sponge contains unique cells called choanocytes. Each of these cells has a long 'tail' called a flagellum. When these tails wiggle, it creates a current that moves the water around inside the sponge's body. Each flagellum has a collar that traps particles for the cells to feed on.

CRAB TAXI
Some crabs pick up sponges and stick them to their shell. This helps the crab to camouflage itself and the sponge gets a free ride. Everyone's a winner!

These tiny creatures are only a few millimetres long, but don't let their size fool you: they are fearsome predators.

4

The hydra can regenerate itself. It gets its name from a monster in Greek mythology with lots of heads. When one head was chopped off, it grew two new ones.

BROWN HYDRA
Hydra oligactis

Animal: **INVERTEBRATE** | Digestive system: **INCOMPLETE** | Type of diet: **CARNIVORE**

Tentacle tactics

Hydrae live in shallow freshwater places, such as ponds, streams and lake shores. They produce a sticky liquid on the underside of their body, and use this to stick themselves to rocks or water plants. Their venomous tentacles are on the upper part of their body. When hunting, they use these tentacles to paralyse, capture and pull their prey into their combined mouth/anus. After digesting it for a couple of days, they expel the remains of their prey through the same opening.

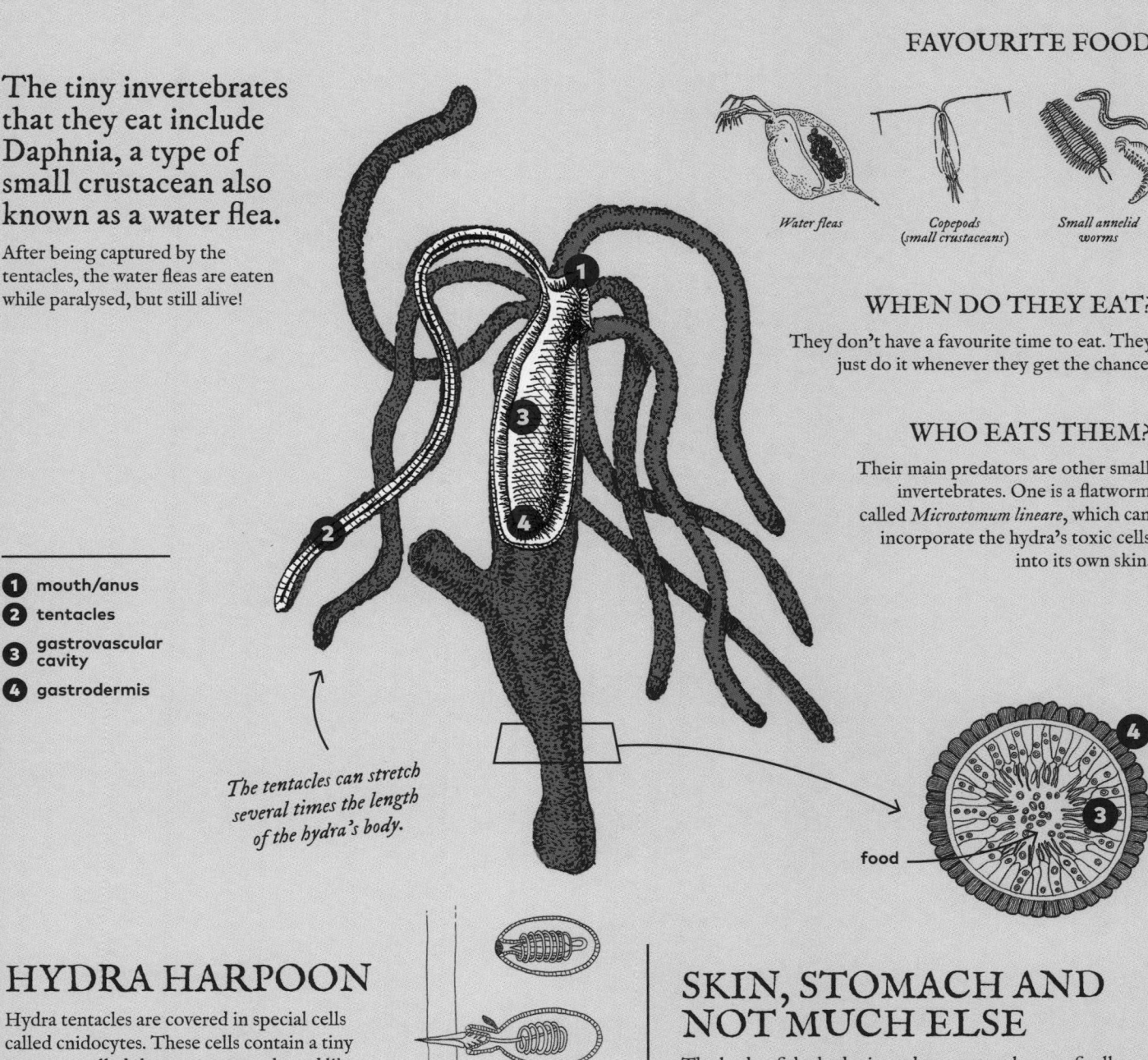

The tiny invertebrates that they eat include Daphnia, a type of small crustacean also known as a water flea.

After being captured by the tentacles, the water fleas are eaten while paralysed, but still alive!

1. mouth/anus
2. tentacles
3. gastrovascular cavity
4. gastrodermis

The tentacles can stretch several times the length of the hydra's body.

FAVOURITE FOOD

Water fleas   *Copepods (small crustaceans)*   *Small annelid worms*

WHEN DO THEY EAT?

They don't have a favourite time to eat. They just do it whenever they get the chance.

WHO EATS THEM?

Their main predators are other small invertebrates. One is a flatworm called *Microstomum lineare*, which can incorporate the hydra's toxic cells into its own skin.

food

HYDRA HARPOON

Hydra tentacles are covered in special cells called cnidocytes. These cells contain a tiny structure called the nematocyst, shaped like a light bulb. When the tentacle touches the hydra's prey, the nematocyst shoots out a harpoon-like thread. This injects toxins that paralyse its prey.

SKIN, STOMACH AND NOT MUCH ELSE

The body of the hydra is made up to two layers of cells separated by thin jelly-like tissue. The epidermis is the outside 'skin' layer, while the gastrodermis is the inner layer. This layer is made of cells that can digest food that's been broken down in the gastrovascular cavity.

The tapeworm is a type of parasite that can grow to a length of 15 metres.

5

There are 32 species of tapeworm in total. Only three species can live inside human beings.

TAPEWORM
Taenia

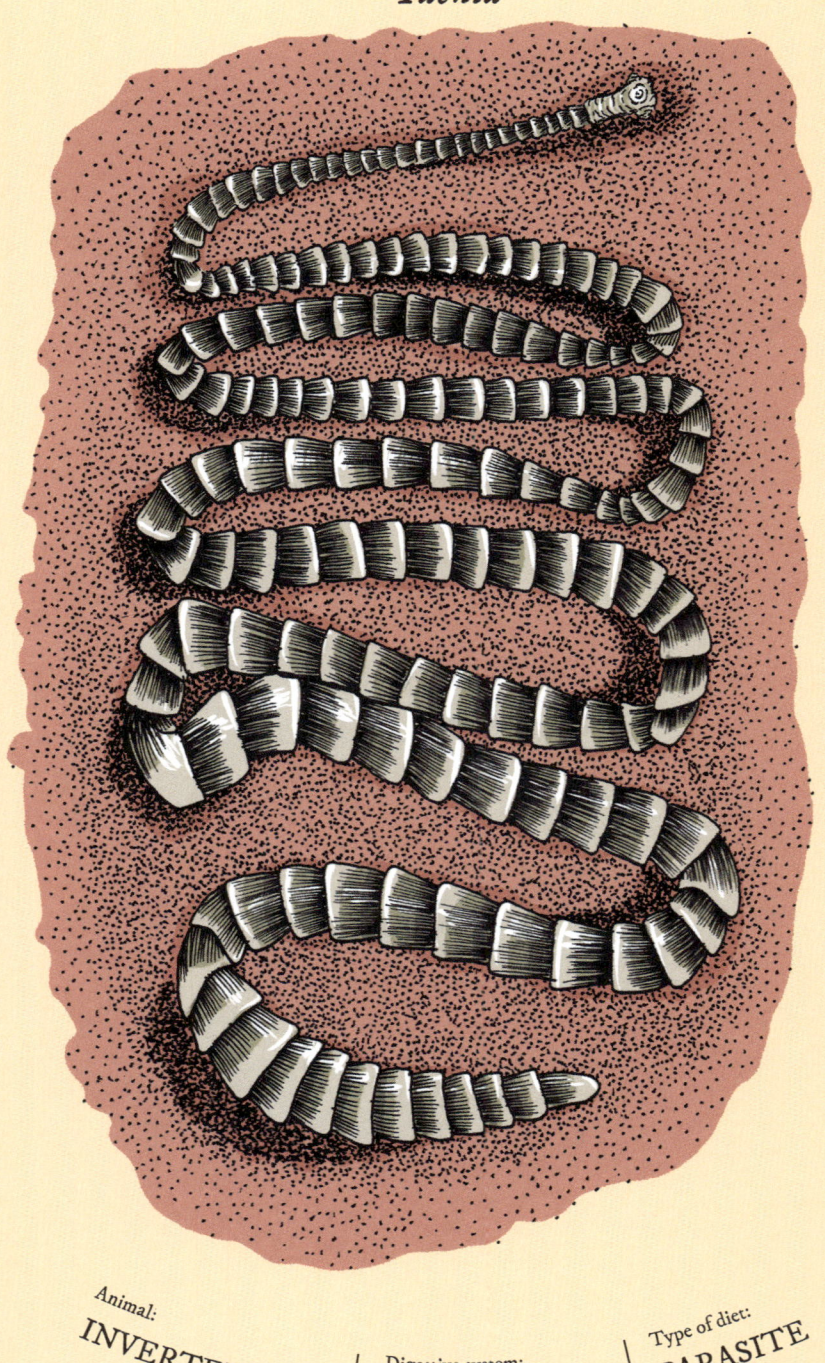

Animal: **INVERTEBRATE** | Digestive system: **INCOMPLETE** | Type of diet: **PARASITE**

No guts, no glory

What better place to find food than in the intestines of another animal? That is where adult tapeworms live, inside the digestive system of their hosts. They could not live in many other places because they don't actually have a digestive tract: no stomach, no intestines and not even a mouth! They absorb nutrients digested by their host, directly through their body.

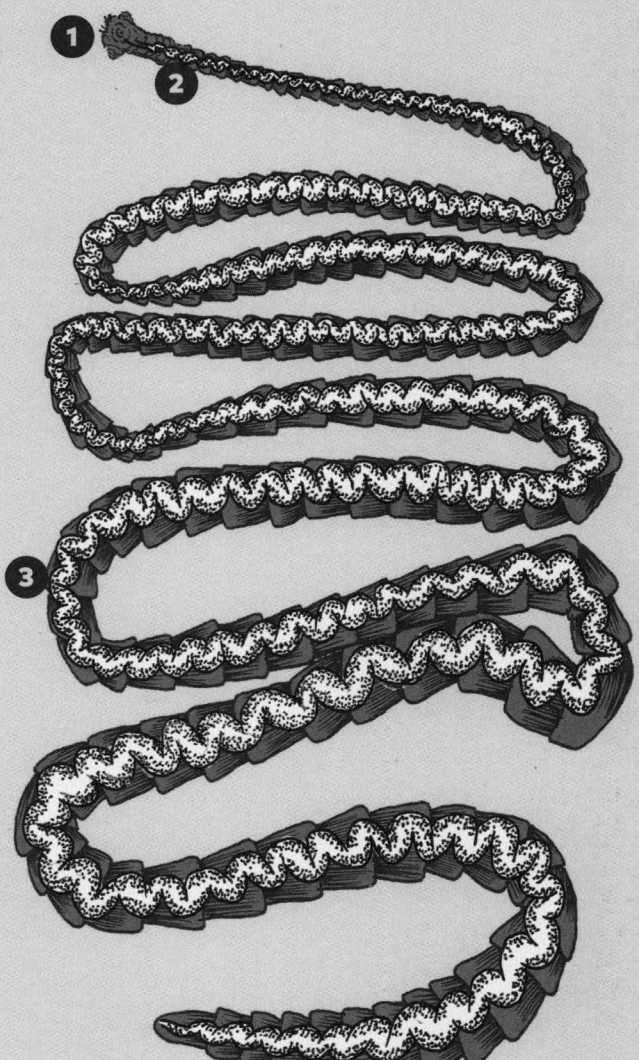

A tapeworm is shaped like a ribbon with three sections: the scolex, the neck and the strobila.

The strobila makes up most of the tapeworm's body and is made up of segments called proglottids. These absorb nutrients from their host.

1. scolex
2. neck
3. strobila (4 to 10 metres long)
4. suckers
5. rostellum

The body of a tapeworm is flat and has no cavities inside it.

FAVOURITE FOOD
Nutrients from their host.

WHEN DO THEY EAT?
The host eats first, digesting when food. When that job is done, the tapeworm gets a free meal!

WHO EATS THEM?
Unlike other animals, tapeworms want to be eaten. They begin life as an egg, growing inside any type of food. Once the host eats the food, they start growing. And it's very hard to get them out!

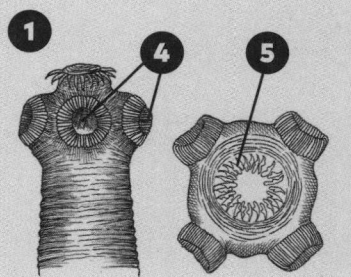

HOLDING ON TIGHT

The scolex is the tapeworm's 'head'. It is covered in suckers that the tapeworm uses to attach itself to the walls of the host's intestine. Some tapeworm species have an extra structure called a rostellum, covered with tiny hooks, which help them to grip.

STOWAWAY

The condition caused by having a tapeworm in your intestines is called taeniasis. Symptoms are very mild, so it is common for a tapeworm to stay undetected in a human until it's grown to a length of several metres. But don't worry! The host can take medicine to get rid of it and then expel it in their poo.

The bodies of jellyfish such as the sea nettle are made up of around 95% water. They live in water and they are made from it too!

Despite popular belief, jellyfish do not just float where the current takes them. They actually swim, and they are very good at it!

SEA NETTLE JELLYFISH
Chrysaora quinquecirrha

Animal:
INVERTEBRATE

Digestive system:
INCOMPLETE

Type of diet:
CARNIVORE

Handle with care!

Jellyfish have been on our planet for more 500 million years. Their digestive system is fairly simple, but it works for them! The secret to their success is their venomous tentacles. Sea nettles have 40 or more tentacles, which they use to kill and eat their prey. Their sting is not deadly to humans, but it can be painful.

FAVOURITE FOOD

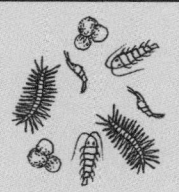
Zooplankton

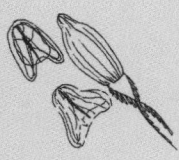
Comb jellies

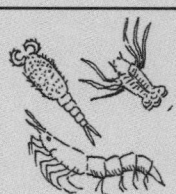
Other jellyfish   *Small crustaceans*

The sea nettle's main source of food is zooplankton – that's plankton made up of tiny animal organisms.

They sometimes eat other sea creatures, but their prey is always very small. So don't worry – although they may sting humans, they certainly won't eat you!

WHEN DO THEY EAT?
As far as we know, these jellyfish are mainly active during the day but they can also move and feed at night.

WHO EATS THEM?
The venom in their tentacles means that they are not very tasty prey, but some predators are immune to their sting. Creatures that eat jellyfish include the leatherback sea turtle, the moonfish, and humans. In some parts of Asia, this jellyfish is considered a delicacy.

1. mouth/anus
2. gastric cavity
3. tentacle
4. oral arm

No need for breaststroke, backstroke or butterfly – nothing beats swimming jellyfish-style!

Sea nettle tentacles can grow to about half a metre in length.

ORAL ARMS
Along with their dozens of tentacles, sea nettles also have four oral arms, which are thicker and slightly shorter than the rest of their tentacles. They use these arms to bring food to their mouth.

SWIMMING SKILLS
Jellyfish can spend more than 90% of the time searching for food, often against the current. They do this by expanding and contracting the top part of their body (the bell) to push themselves through the water. To work out where they're going, they have very simple eyes that help them 'feel' their way towards light.

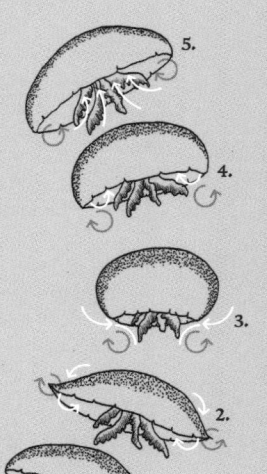

7.
Starfish

8.
Green sea slug

9.
Common spiny lobster

10.
Tarantula

11.
Honey bee

12.
Medicinal leech

13.
Dung beetle

14.
Sea cucumber

15.
Blue sea dragon

16.
Volcano snail

17.
Mosquito

18.
Common octopus

19.
Shipworm

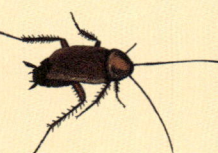

20.
Black cockroach

21.
Pacific sand dollar

22.
Housefly

23.
Termite

24.
Face mite

25.
Roman snail

26.
Black-lip pearl oyster

INVERTEBRATES
Complete digestive system

Starfish look very pretty, but the way they feed is quite terrifying to their prey!

A few species of starfish have an incomplete digestive system with a single mouth/anus. But most have a mouth underneath and a separate anus on top of their body.

STARFISH

Asteroidea

Animal: **INVERTEBRATE** | Digestive system: **COMPLETE** | Type of diet: **CARNIVORE**

Inside out!

Many starfish have a very unusual skill: they can push one of their two stomachs out of their mouth! They use this pushed-out stomach as a net to catch their prey. If the prey has a double shell (like a mussel), they can also use their stomach to force open its shell and will stuff their guts inside it to feed on the soft flesh.

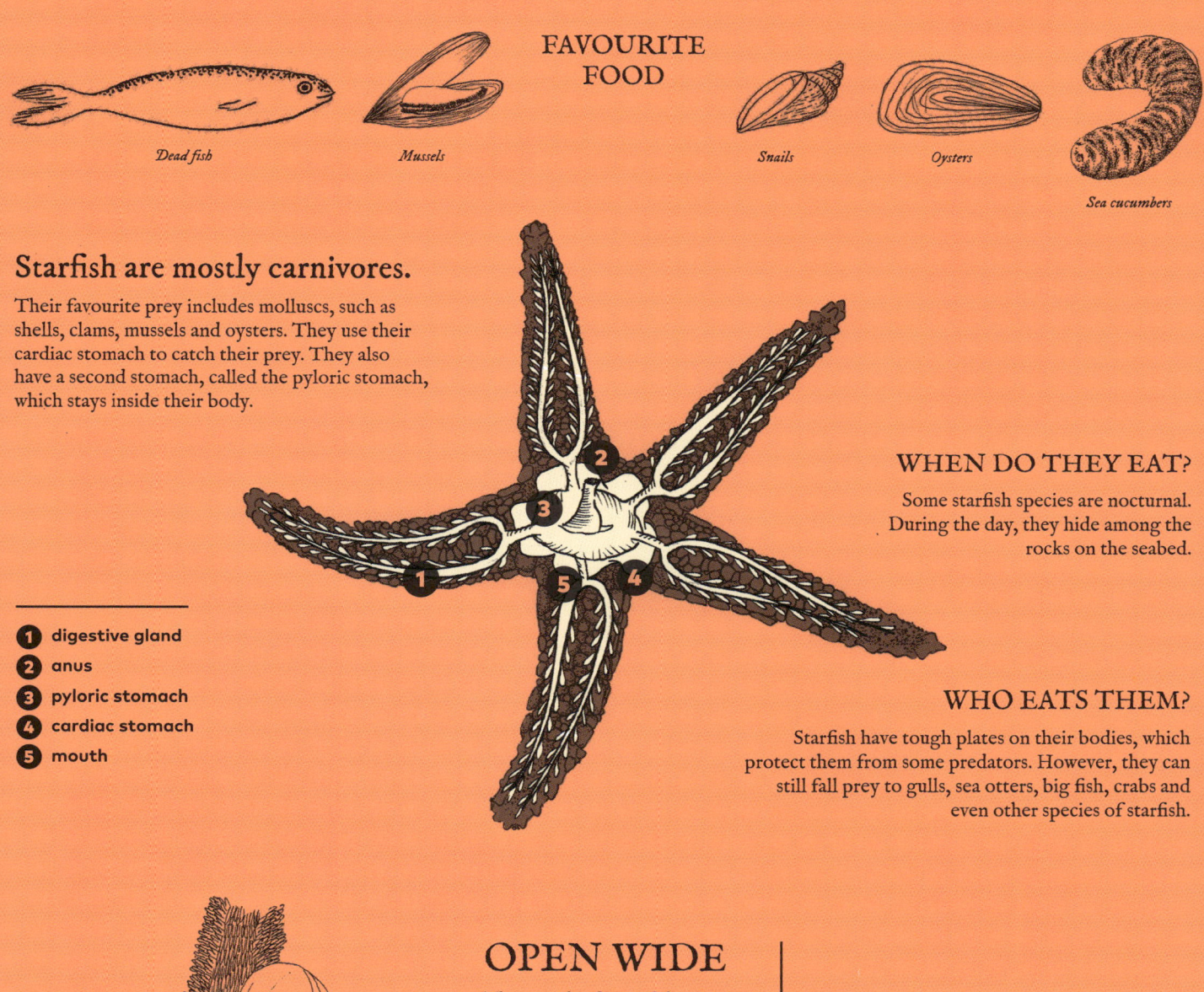

FAVOURITE FOOD

Dead fish — *Mussels* — *Snails* — *Oysters* — *Sea cucumbers*

Starfish are mostly carnivores.

Their favourite prey includes molluscs, such as shells, clams, mussels and oysters. They use their cardiac stomach to catch their prey. They also have a second stomach, called the pyloric stomach, which stays inside their body.

1. digestive gland
2. anus
3. pyloric stomach
4. cardiac stomach
5. mouth

WHEN DO THEY EAT?

Some starfish species are nocturnal. During the day, they hide among the rocks on the seabed.

WHO EATS THEM?

Starfish have tough plates on their bodies, which protect them from some predators. However, they can still fall prey to gulls, sea otters, big fish, crabs and even other species of starfish.

OPEN WIDE

The mouth of a starfish is on the underside of its body, right in the middle.

This is what the inside-out stomach of a starfish looks like

THE POWER TO REGENERATE

If a starfish is injured by a predator but is eaten, it can still recover. Some species can grow a new arm if they lose one. And a lucky few can grow an entire new body from one arm!

This green mollusc not only looks like a plant, but feeds like one too.

8

As adults, these slugs can survive for months without eating, by making nutrients from sunshine.

GREEN SEA SLUG
Elysia chlorotica

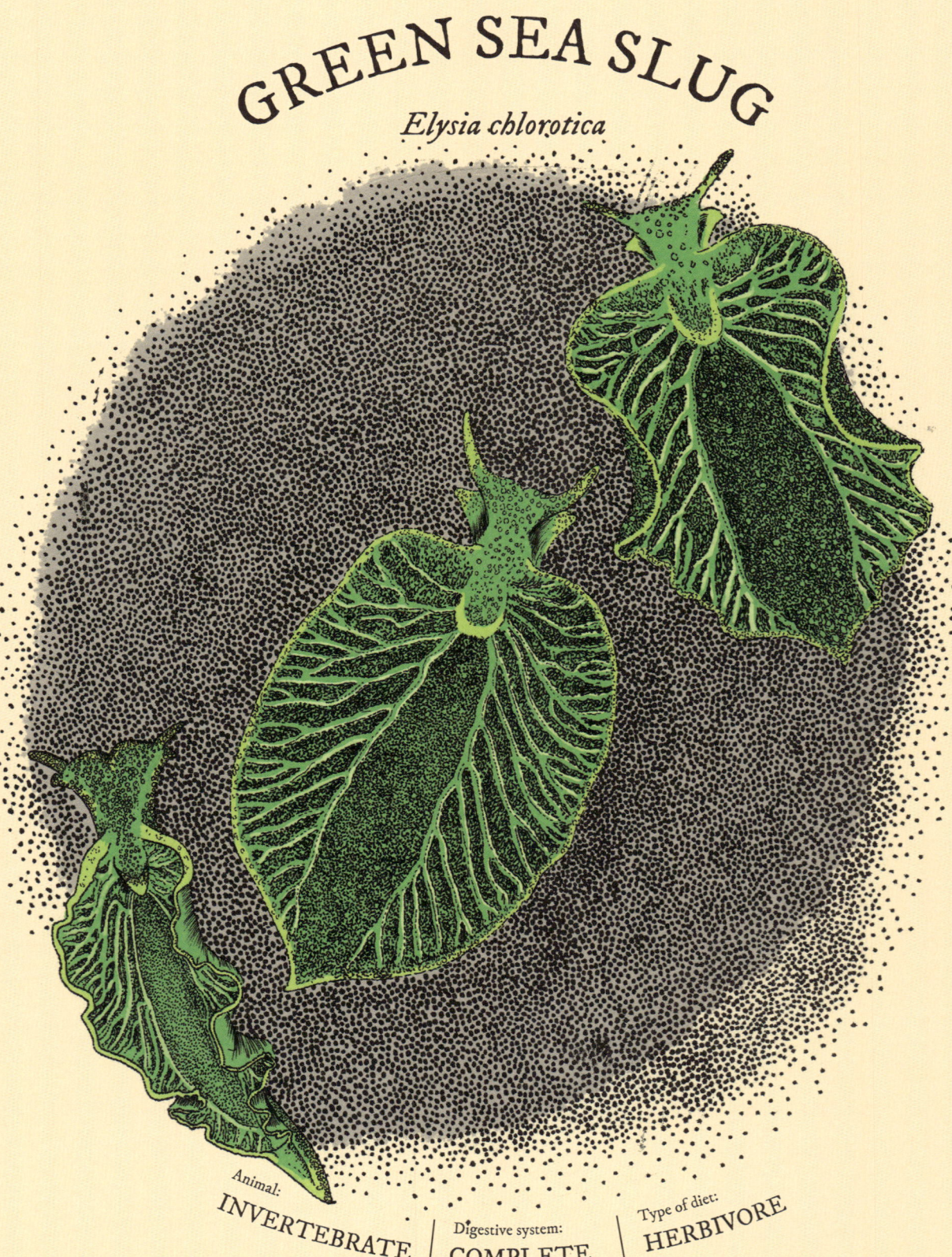

Animal:
INVERTEBRATE

Digestive system:
COMPLETE

Type of diet:
HERBIVORE

Eat like a leaf

Unlike animals, plants get their energy through a process called photosynthesis. This allows them to use sunlight as 'fuel' to turn water and a gas called carbon dioxide into the nutrients they need. The green sea slug is one of the few animals that can copy this technique. It does it by taking cells from the algae it eats and using them in its digestive process.

FAVOURITE FOOD

Vaucheria litorea

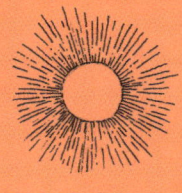
Sunlight

The diet of these sea slugs never changes. They feed on only one type of algae, *Vaucheria litorea*.

When a slug digests this algae, it absorbs some of its cells, called chloroplasts, into its own stomach cells. The chloroplasts contain a substance called chlorophyll, which plants use to capture energy from sunlight. With the chloroplasts in its stomach, the slug gains that skill too!

WHEN DO THEY EAT?

These animals are active during the day because they need sunlight to get their energy.

WHO EATS THEM?

The green sea slug has no known predators. Its green colour acts as camouflage when it hides in the seaweed.

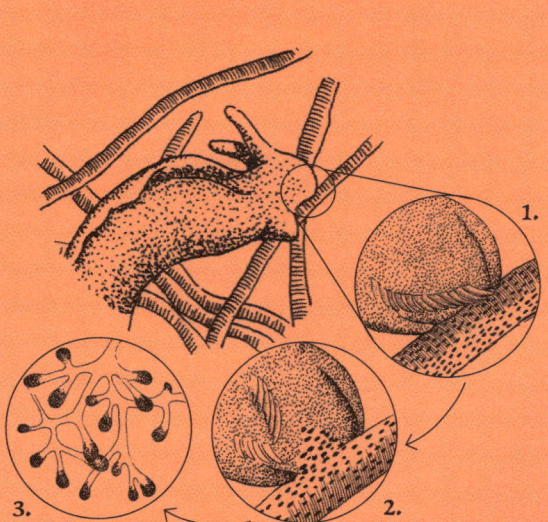

1 radula (teeth)
2 anus
3 digestive organs
4 dorsal veins

SEAWEED SMOOTHIE

The radula is what slugs and snails have instead of a tongue and teeth. It is like a tongue covered in tiny teeth. The sea slug uses it to scrape off the outer layer of algae. Then it sucks out the inside, like drinking a smoothie through a straw.

GOING GREEN

Before they get their first taste of chlorophyll, these sea slugs are brown with red spots. When they begin feeding, chloroplast cells from the algae they eat become part of their bodies, and they turn bright green.

When food is scarce, these animals sometimes resort to cannibalism and eat other lobsters, especially smaller ones.

It's not only other lobsters that eat lobsters — they are also one of the most popular seafoods for humans to eat.

COMMON SPINY LOBSTER

Palinurus elephas

Animal:
INVERTEBRATE

Digestive system:
COMPLETE

Type of diet:
OMNIVORE

A stomach with teeth!

Lobsters have two stomachs. The first is called the cardiac sac, which contains a tough structure called the gastric mill. This is used to grind up the lobster's food — it's as if the lobster had teeth in its stomach! When the food is mashed up, it passes through to the second stomach, which is called the pyloric sac.

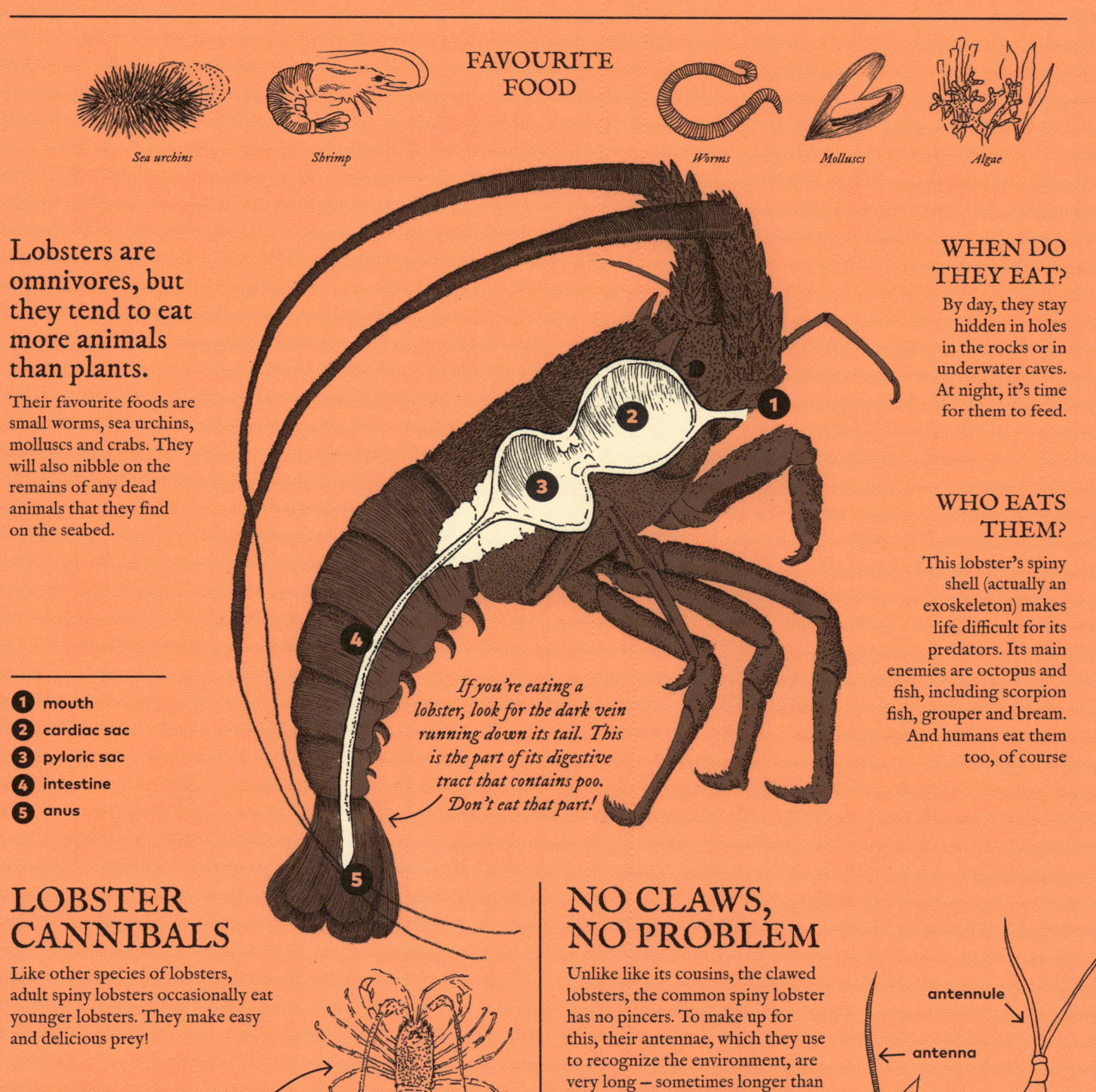

FAVOURITE FOOD

Sea urchins — *Shrimp* — *Worms* — *Molluscs* — *Algae*

Lobsters are omnivores, but they tend to eat more animals than plants.

Their favourite foods are small worms, sea urchins, molluscs and crabs. They will also nibble on the remains of any dead animals that they find on the seabed.

1. mouth
2. cardiac sac
3. pyloric sac
4. intestine
5. anus

WHEN DO THEY EAT?

By day, they stay hidden in holes in the rocks or in underwater caves. At night, it's time for them to feed.

WHO EATS THEM?

This lobster's spiny shell (actually an exoskeleton) makes life difficult for its predators. Its main enemies are octopus and fish, including scorpion fish, grouper and bream. And humans eat them too, of course

If you're eating a lobster, look for the dark vein running down its tail. This is the part of its digestive tract that contains poo. Don't eat that part!

LOBSTER CANNIBALS

Like other species of lobsters, adult spiny lobsters occasionally eat younger lobsters. They make easy and delicious prey!

Lobsters are happy to eat the children of other lobsters!

NO CLAWS, NO PROBLEM

Unlike like its cousins, the clawed lobsters, the common spiny lobster has no pincers. To make up for this, their antennae, which they use to recognize the environment, are very long — sometimes longer than their whole body. They also have four shorter antennules, which they use to 'smell' food.

antennule
antenna

Unlike most other spiders, tarantulas don't use webs to catch their prey. They spin silk only to build dens or as a way to detect movement.

Tarantulas can survive a long time without food. Some species, like the Chilean rose tarantula, can live for more than two years without eating anything.

TARANTULA
Theraphosidae

Animal:
INVERTEBRATE

Digestive system:
COMPLETE

Type of diet:
CARNIVORE

A natural liquidizer

Tarantulas can only eat tiny particles, measuring one micrometre (equivalent to one thousandth of a millimetre) at the most. So before eating their prey, they need to turn it into liquid. This is called external digestion. It means the digestive process begins before they even have their first mouthful!

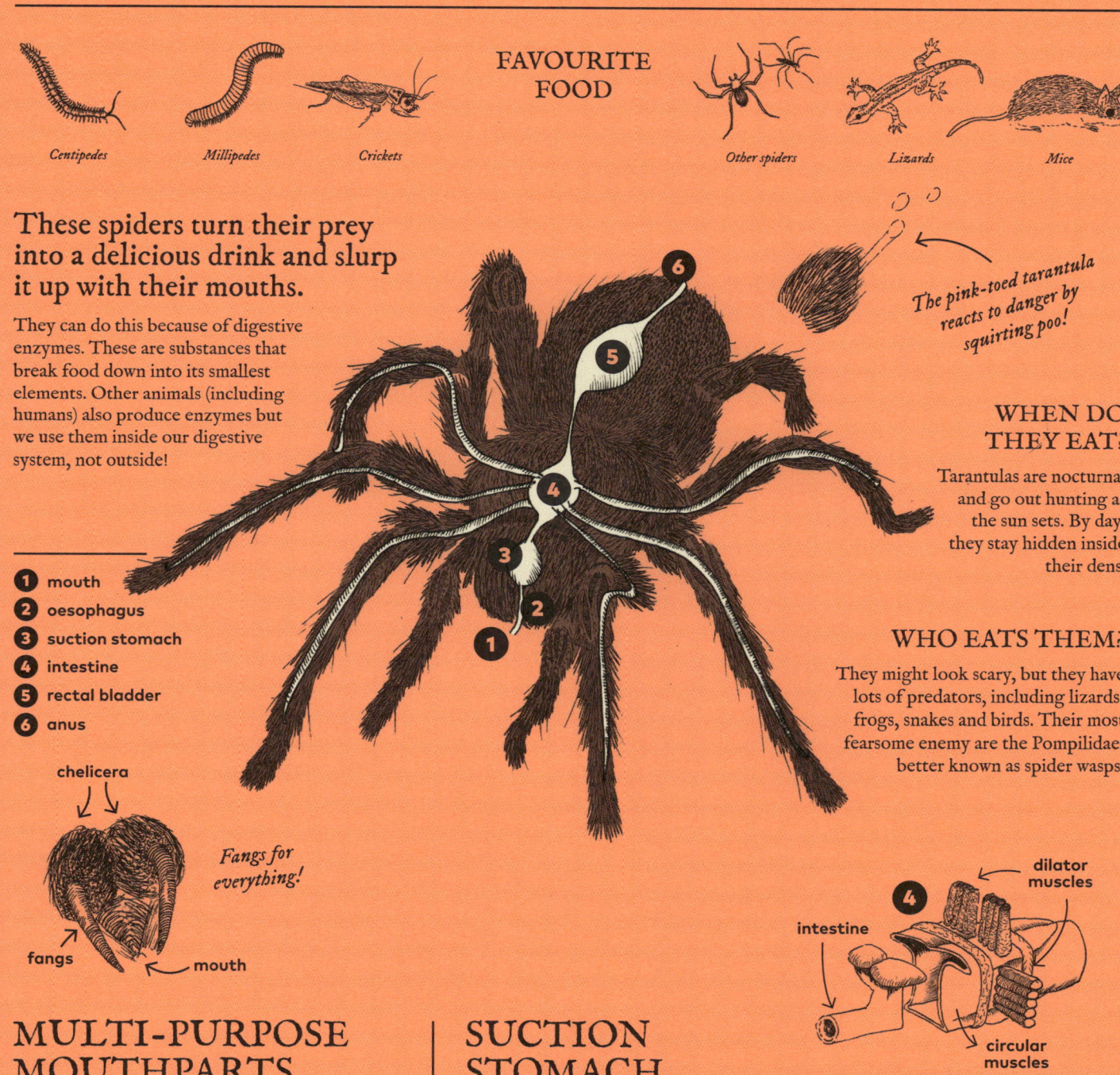

FAVOURITE FOOD
Centipedes · Millipedes · Crickets · Other spiders · Lizards · Mice

These spiders turn their prey into a delicious drink and slurp it up with their mouths.

They can do this because of digestive enzymes. These are substances that break food down into its smallest elements. Other animals (including humans) also produce enzymes but we use them inside our digestive system, not outside!

1. mouth
2. oesophagus
3. suction stomach
4. intestine
5. rectal bladder
6. anus

Fangs for everything!

chelicera · fangs · mouth

The pink-toed tarantula reacts to danger by squirting poo!

WHEN DO THEY EAT?
Tarantulas are nocturnal and go out hunting as the sun sets. By day, they stay hidden inside their dens.

WHO EATS THEM?
They might look scary, but they have lots of predators, including lizards, frogs, snakes and birds. Their most fearsome enemy are the Pompilidae, better known as spider wasps.

4. intestine · dilator muscles · circular muscles

MULTI-PURPOSE MOUTHPARTS

The mouthparts of a tarantula are called chelicerae. They use them to inject venom into their prey and kill it. They also produce powerful saliva, full of digestive enzymes that will turn their prey to liquid.

SUCTION STOMACH

In the front of the spider's body, there's an organ called the suction stomach. This is one of the tarantula's digestive organs. Its strong muscles act like a pump, helping the spider to slurp up its liquidized prey.

There are three types of bees in a colony: the queen who lays eggs, the drones who are male, and the workers who hunt for food and build honeycombs.

Bees have an amazing sense of smell. They can detect flowers from up to two kilometres away.

HONEY BEE
Apis mellifera

Animal: **INVERTEBRATE** | Digestive system: **COMPLETE** | Type of diet: **HERBIVORE**

Sweet as honey

The most important digestive organ for worker bees is the crop or honey sac. This is where they store the nectar and water they collect, before carrying it back to the hive to be turned into honey. The next section of their digestive system is the true stomach or ventriculus, where the bee digests its own food.

FAVOURITE FOOD

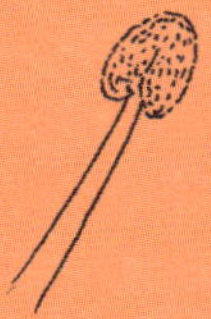

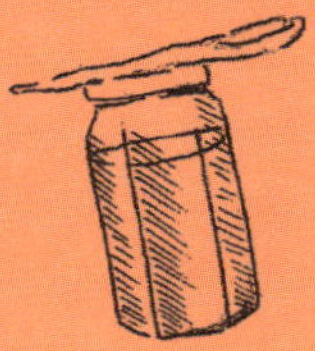

Nectar   *Pollen*   *Honey* 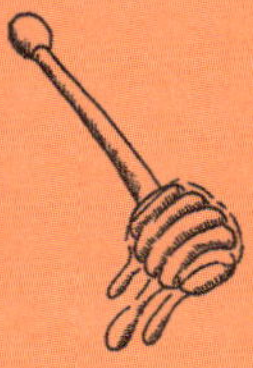   *Royal jelly* 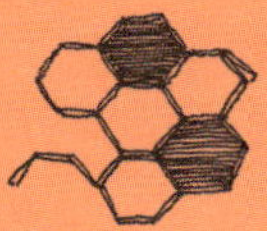   *Bee bread*

The two main foods in a bee's diet are nectar and pollen.

They use the nectar to make honey, bee bread (a mixture of pollen and nectar), and royal jelly. They eat the first two of these when food is scarce. Royal jelly is only eaten by bee larvae and the queen bee.

1. proboscis
2. oesophagus
3. honey sac (crop)
4. intestine
5. anus

WHEN DO THEY EAT?
Worker bees collect nectar and pollen during the day, especially in the hotter months. In the winter, they feed on honey from the hive.

WHO EATS THEM?
Honey and larvae from the hives are eaten by bears, honey badgers, skunks and raccoons. Wasps, spiders and birds such as bee-eaters also prey on bees.

Bees don't poo inside the hive, not even during the long winter months. Instead, waste builds up inside their digestive system. Then when the time is right, they go outside to get rid of the poo and to clean themselves.

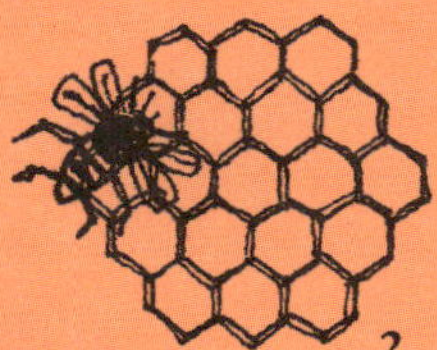

MAKING HONEY

The process begins with the workers going from flower to flower. From each flower, they collect nectar, a sugary liquid, and store it in their crop or honey sac. Then they return to the hive and regurgitate the nectar through their mouths. Other younger workers add chemicals called enzymes, which turn the nectar into honey. The honey is stored in the cells of the honeycomb.

A NATURAL STRAW

Bees suck up nectar with a long tube-like organ called a proboscis.

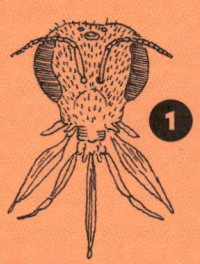

The leeches of this species are haematophages. This means that they feed on the blood of other animals.

Historically, these leeches were used by human doctors to treat their patients. This is why they are known as medicinal leeches.

MEDICINAL LEECH
Hirudo medicinalis

Animal:
INVERTEBRATE | Digestive system: **COMPLETE** | Type of diet: **HAEMATOPHAGE**

Bloodsuckers!

To feed, leeches stick themselves to their prey, using the suckers on their mouth and anus. When they are fastened on tight, they bite a hole in the animal's skin and use their powerful pharynx (throat) to suck up blood. A leech can eat more than five times its own weight in blood! Its body swells up and it won't need to eat again for at least six months.

FAVOURITE FOOD

Blood

These bloodsuckers like to drink from any mammals who enter the ponds, swamps or streams where they live. This includes humans!

However, young leeches prefer to prey on frogs, which have thinner skin. This is easier on their jaws, which haven't grown strong yet.

WHEN DO THEY EAT?

They are normally more active in the evening, but they don't mind drinking blood during the day.

WHO EATS THEM?

They have quite a few predators, including water birds (such as ducks), fish, snakes and even some frogs.

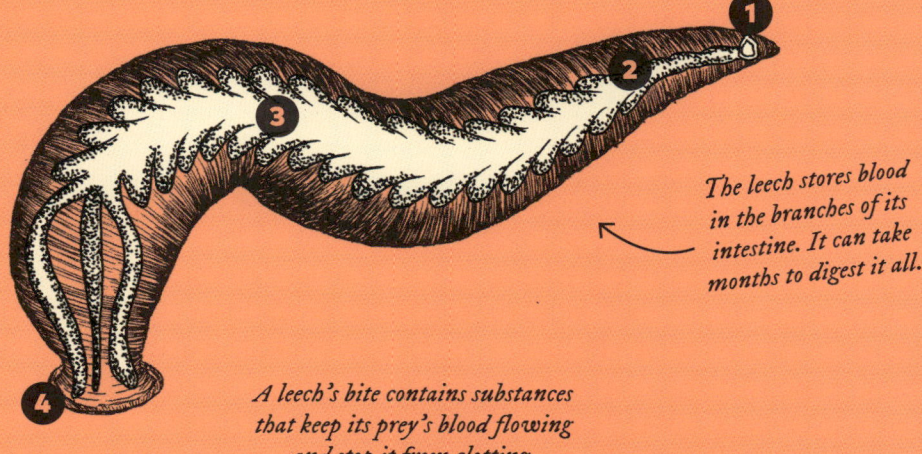

1. anterior sucker/jaws
2. crop
3. intestine
4. posterior sucker

The leech stores blood in the branches of its intestine. It can take months to digest it all.

A leech's bite contains substances that keep its prey's blood flowing and stop it from clotting.

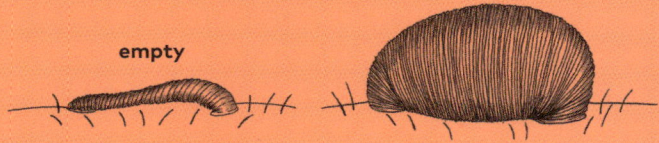

empty — full

TRIPLE JAWS

Leeches have three jaws, each with about a hundred tiny teeth. But the leech's prey often doesn't notice that it's been bitten. The leech has anaesthetic in its saliva, so the bite doesn't hurt.

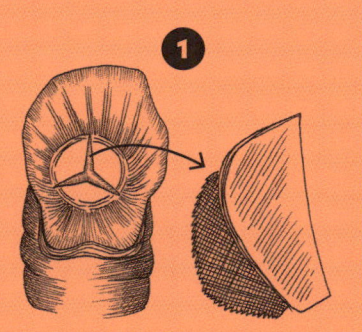

A MEDICAL MARVEL

Historically, people believed that some illnesses could be cured with blood-letting, which meant extracting 'bad' blood from the patient. Leeches were used to do this. They are still used for some very specific medical purposes. The substances in their saliva are also being studied because they could be very useful for humans.

Most beetles in the Scarabaeinae subfamily feed on animal dung. But they never complain about what's for dinner!

A dung beetle can eat more than its own weight in poo in only one day.

DUNG BEETLE

Scarabaeinae

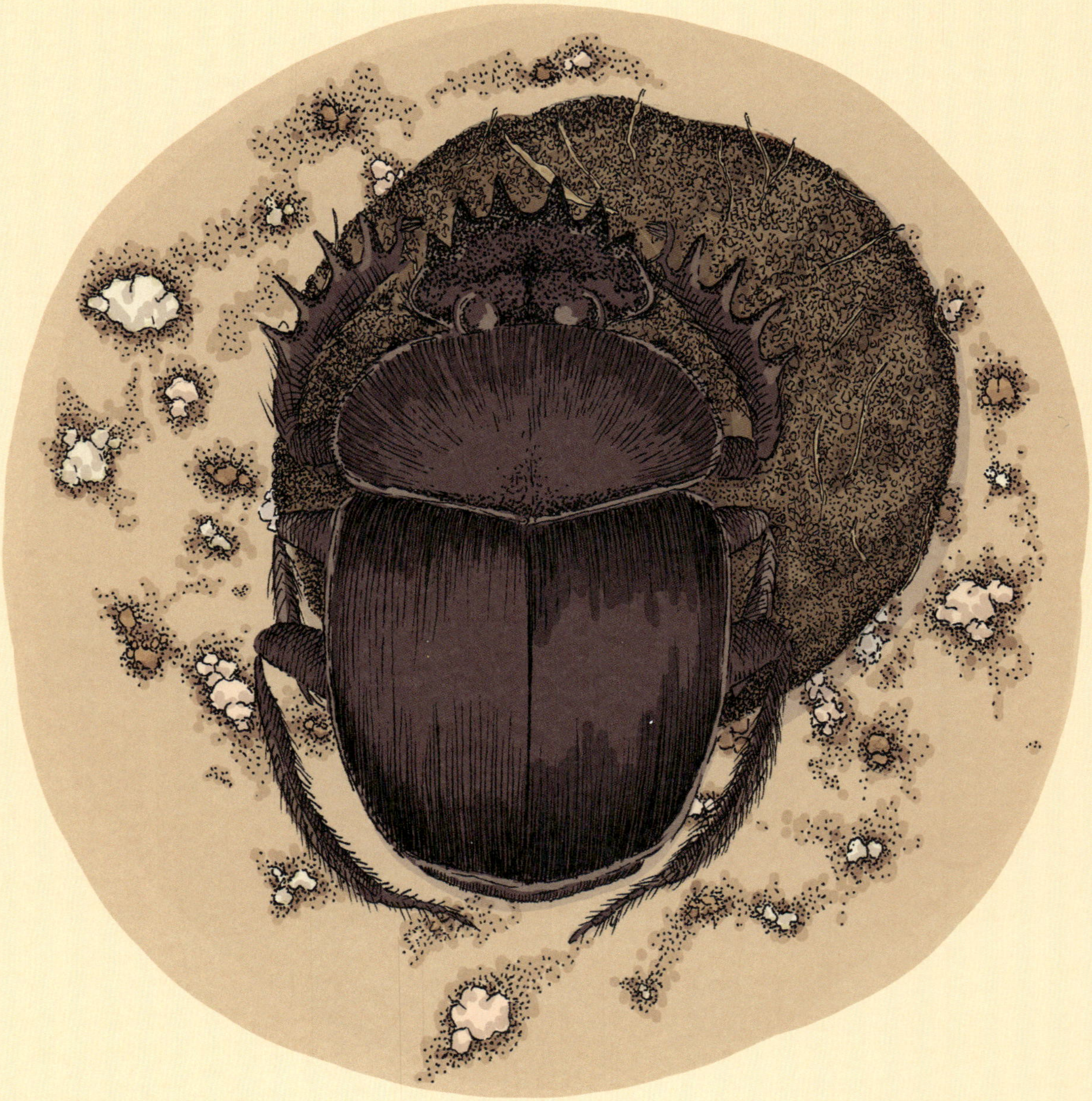

Animal: **INVERTEBRATE** | Digestive system: **COMPLETE** | Type of diet: **COPROPHAGE**

Rolling, rolling, rolling

Adult dung beetles feed mostly on the liquid part of animal droppings. It's full of microorganisms and they think it's delicious! Then they gather up the dry parts of the droppings and roll them into a ball. To help them do this, they have a shovel-shaped head and flattened front legs with tooth-like prongs. They roll the dung ball to the underground tunnels where they live, and feed it to their larvae.

Nearly all dung beetles prefer to eat the dung of herbivores or omnivores, because it contains more fibre than carnivore dung. Some species also eat carrion meat or rotting organic matter.

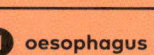

1. oesophagus
2. crop
3. stomach
4. intestine

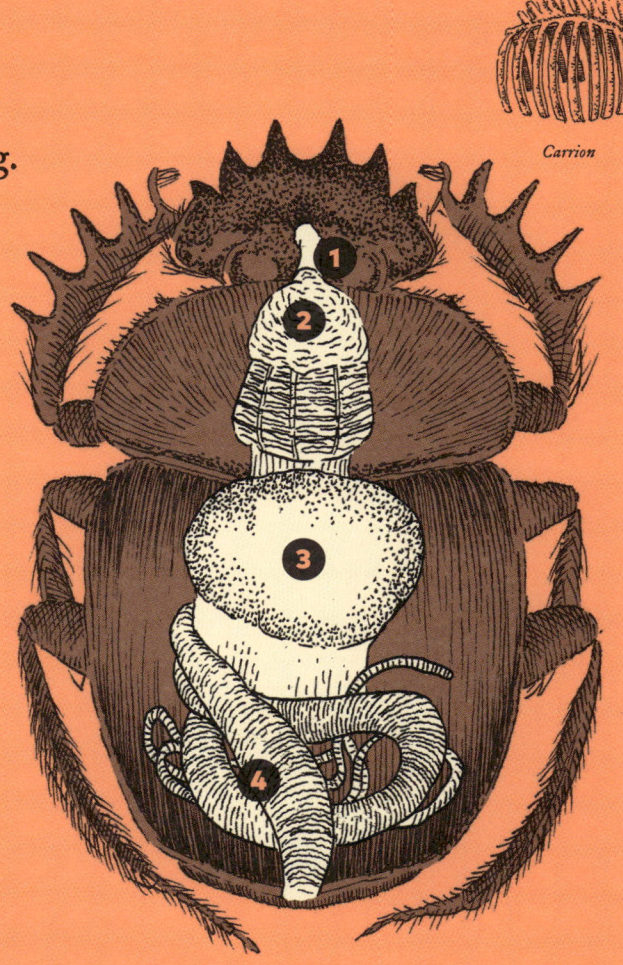

FAVOURITE FOOD

Carrion   *Dung*   *Rotten fruit and fungi*

WHEN DO THEY EAT?
Some species (especially in colder places) are active during the day, while others (especially in warmer places) are most active at dusk or during the night.

WHO EATS THEM?
Their main predators are insect-eating birds. They are also preyed on by other insects, such as wasps and ants.

Some species make dung balls that are almost as large as a tennis ball.

HOME IS WHERE THE DUNG IS

Adult beetles bring balls of dung back to their tunnels and turn them into smaller balls. Female beetles lay an egg in each small ball. This is where the dung beetle larvae are born and grow. The dung keeps them warm and means they have plenty of food until they are strong enough to come out from their little poo home.

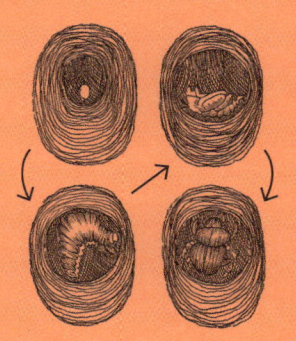

ECO-BEETLES

They may sound disgusting but the truth is that dung beetles are one of the most eco-friendly creatures in the world. 'Recycling' poo makes the soil they live in more fertile. It also reduces the number of parasites that can harm livestock.

Creatures from the Holothuroidea family are commonly known as sea cucumbers because of their long shape. They live in oceans all over the world.

As you might guess, they are not the fastest animals in the ocean, but they can move by crawling on their very tiny feet, or in some cases, by swimming.

SEA CUCUMBER
Holothuroidea

Animal: **INVERTEBRATE** | Digestive system: **COMPLETE** | Type of diet: **DETRITIVORE**

Your stomach or your life!

The digestive system of a sea cucumber is not only used for feeding. It can also be used as a clever tactic when a predator attacks them. In these situations, they can push their guts right out of their body (through the mouth or the anus, depending on the species). This startles and distracts their attacker while the cucumber escapes. A sea cucumber can survive without its intestines for long enough to grow new ones.

FAVOURITE FOOD

Organic remains

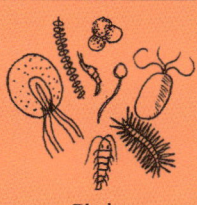
Plankton

Their main source of food is detritus, which means rotting bits of plant or animal matter.

Using the tentacles around their mouth, they catch these particles as they float through the water or lie on the seafloor.

WHEN DO THEY EAT?

Like most detritivores who live in the sea, they eat at any time.

WHO EATS THEM?

Despite their defence mechanisms, sea cucumbers are prey for starfish, fish, snails and crustaceans. Humans also use them as food. They are considered a delicacy in some parts of Asia.

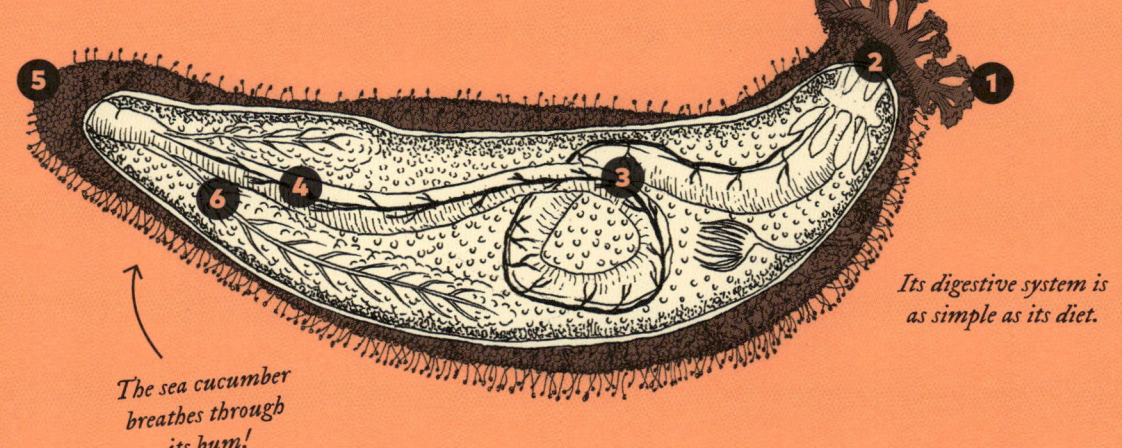

Its digestive system is as simple as its diet.

The sea cucumber breathes through its bum!

1. tentacles
2. mouth
3. stomach
4. intestine
5. anus
6. respiratory tree

TENTACLES AT THE FRONT...

The sea cucumber's mouth is surrounded by short tentacles. These vary in shape, depending on the species. Some are shaped like fingers, others like feathers, and others are flat. These organs also produce sticky mucus that helps the cucumbers to catch food.

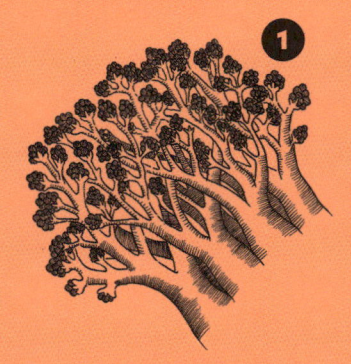

... AND TREES AT THE BACK

Respiratory trees are branch-like structures that the sea cucumber uses to breathe, instead of lungs or gills. They are attached to the cucumber's intestine, alongside long thread-like tubes called Cuvierian tubules. Some species can push these outside their bodies to scare away predators.

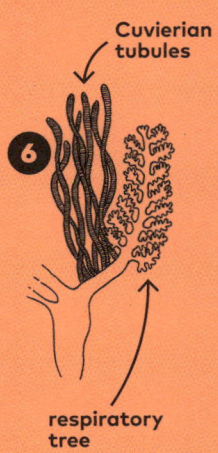

Cuvierian tubules

respiratory tree

Although its colours are beautiful, this species of sea slug is one of nature's most dangerous animals for its size.

15

What you see here is not the top surface but the underside of its body. The blue sea dragon actually spends its life floating upside down!

BLUE SEA DRAGON
Glaucus atlanticus

Animal:
INVERTEBRATE

Digestive system:
COMPLETE

Type of diet:
CARNIVORE

Pretty toxic!

What makes this pretty little creature especially dangerous is its diet. It likes to feed on venomous creatures such as the Portuguese man-of-war, a false jellyfish with a deadly sting. The venom cells of the man-of-war pass through the digestive tract of the slug, but are not digested. Instead they are stored inside the slug's body. The slug can then use this venom to sting its predators.

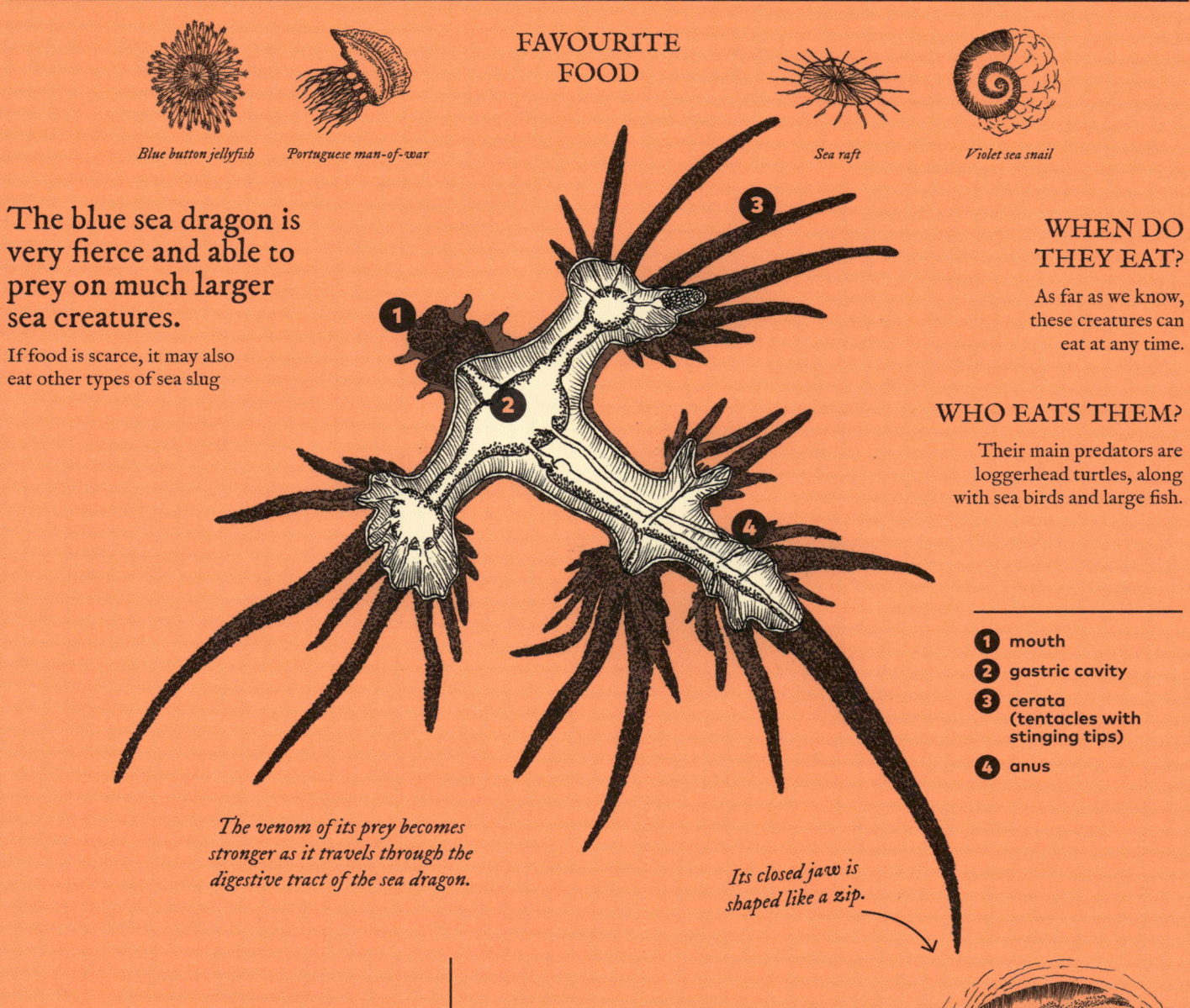

FAVOURITE FOOD

Blue button jellyfish *Portuguese man-of-war* *Sea raft* *Violet sea snail*

The blue sea dragon is very fierce and able to prey on much larger sea creatures.

If food is scarce, it may also eat other types of sea slug

WHEN DO THEY EAT?
As far as we know, these creatures can eat at any time.

WHO EATS THEM?
Their main predators are loggerhead turtles, along with sea birds and large fish.

1. mouth
2. gastric cavity
3. cerata (tentacles with stinging tips)
4. anus

The venom of its prey becomes stronger as it travels through the digestive tract of the sea dragon.

Its closed jaw is shaped like a zip.

BELLY UP!
The blue sea dragon floats with the help of an air sac in its stomach. It swims upside down in order to stay camouflaged. If predators look down on it from above, its blue belly blends in with the water. If they look up at it from below, its silvery top side merges with the surface of the sea.

SNIFF AND SNAP
To find their prey, sea dragons use sensors called rhinophores, which are the equivalent to our sense of smell. When they get close their prey, their powerful jaws snap into action. Their razor-sharp teeth allow them to grab hold of almost anything.

Also known as the scaly-foot gastropod, this snail makes its home close to undersea volcanoes.

It lives 2,500 to 3,000 metres below the sea, close to hydrothermal vents. These are cracks in the sea bed, gushing with boiling hot water that's heated by the volcanic activity below.

VOLCANO SNAIL
Chrysomallon squamiferum

Animal: **INVERTEBRATE** | Digestive system: **COMPLETE** | Type of diet: **CHEMOSYNTHESIS**

The no-food diet

One of the most fascinating facts about this snail is that it doesn't need to eat to stay alive. The volcano snail survives because of bacteria that it carries in a gland in its oesophagus. These friendly bacteria carry out a process called chemosynthesis. It's similar to photosynthesis in plants but instead of using sunlight, the bacteria turn chemicals from the hydrothermal vents into energy.

It could be said that this snail feeds on sulphur, although it does not actually digest it.

The chemosynthetic bacteria that live in its digestive system turn the chemicals produced by volcanoes into the nutrients that the snail needs to stay alive.

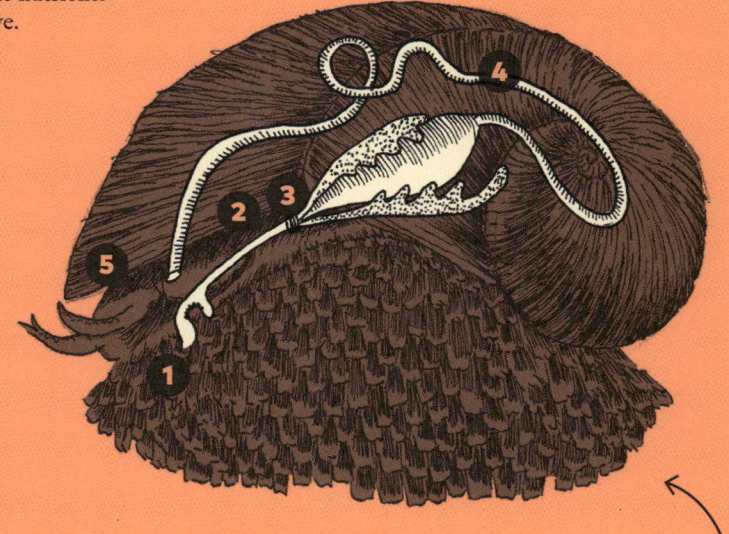

1. mouth
2. oesophageal gland
3. digestive gland
4. intestine
5. anus

Grains of sulphur form inside the intestines of these snails. This is thought to help their unusual digestive system.

FAVOURITE FOOD
None!

WHEN DO THEY EAT?
They don't! But the process of chemosynthesis never stops.

WHO EATS THEM?
Their main threat is other sea snails from the *Phymorhynchus* genus, but the volcano snail's shell is good protection against them!

A STORE OF BACTERIA

The oesophageal gland is an organ found in some snails and slugs. In the case of the volcano snail, it holds the bacteria that provide it with nutrients, so it is especially important. This is reflected in its size: it makes up about 10% of the snail's body!

MADE OF METAL

The shell of the volcano snail is unique in nature. Its inner layer is like that of other snails and slugs. Next comes a layer of soft tissue that cushions blows. The outer layer is the most fascinating: it contains iron sulphides, so it's basically a suit of armour!

Iron sulphides
Organic tissue
Calcium carbonate

This pesky insect causes more human deaths than any other creature on Earth. It does this by passing on diseases through its bite.

17

The life of a mosquito is extremely short: one week for males and even less for females.

MOSQUITO
Culicidae

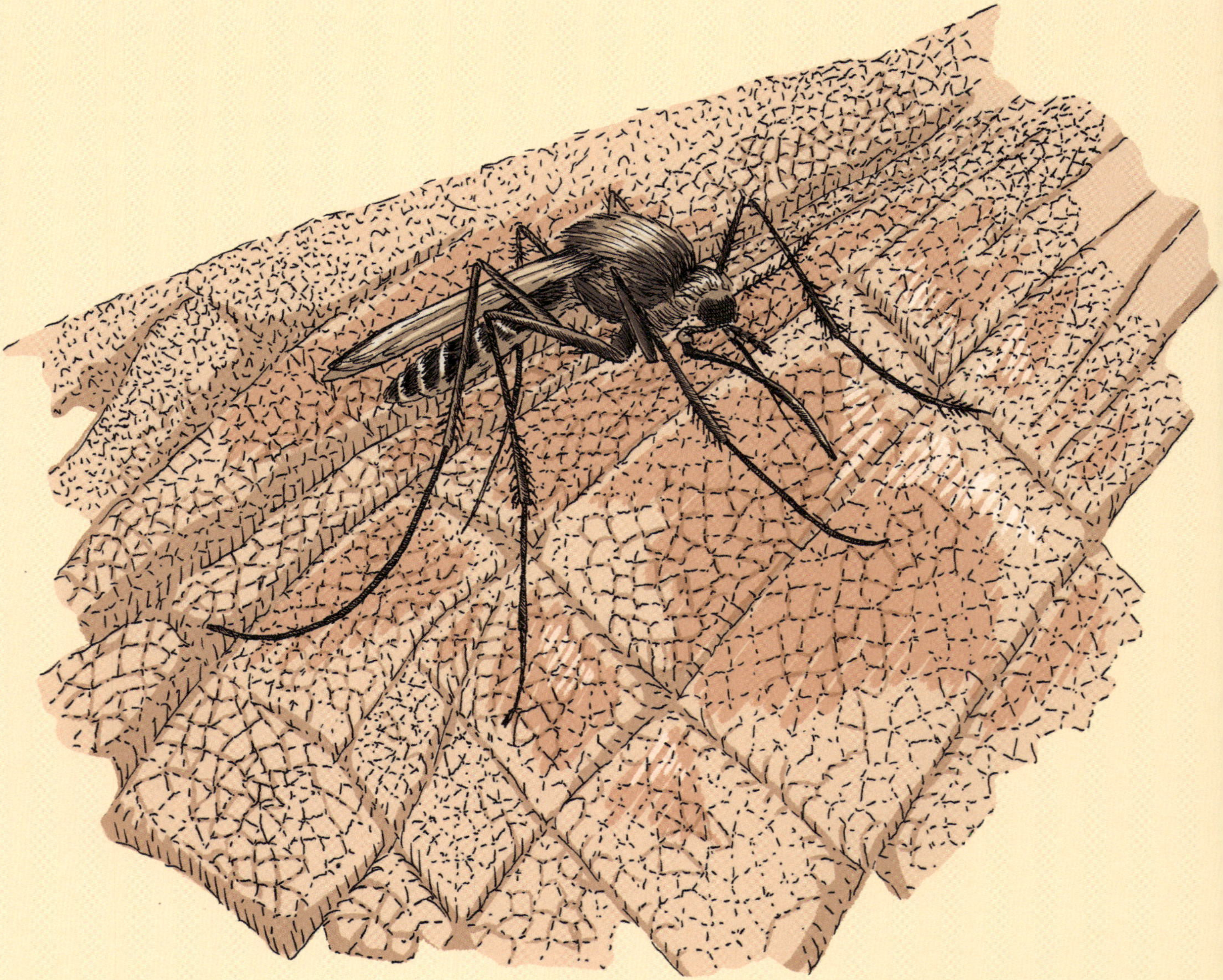

Animal:
INVERTEBRATE

Digestive system:
COMPLETE

Type of diet:
HAEMATOPHAGE

Mosquitoes and miss-quitoes

Mosquitoes are known for drinking blood but in fact not every mosquito does this. The only ones that follow a vampire diet are female mosquitos. For some species, the substances (especially protein) found in blood are needed in order to lay eggs. For other species, the substances mean they can lay more eggs, meaning that there's more chance of their children surviving.

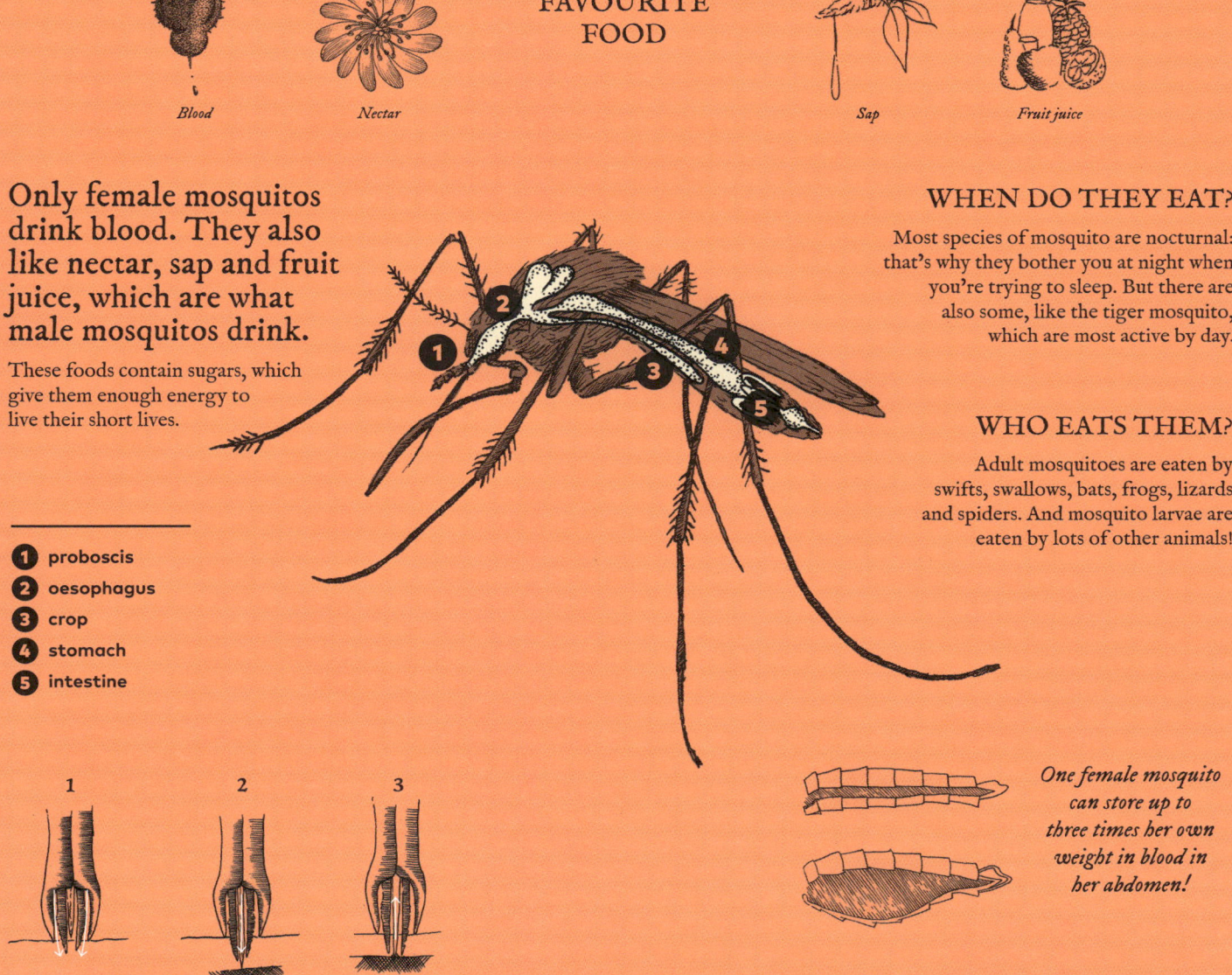

FAVOURITE FOOD

Blood — *Nectar* — *Sap* — *Fruit juice*

Only female mosquitos drink blood. They also like nectar, sap and fruit juice, which are what male mosquitos drink.

These foods contain sugars, which give them enough energy to live their short lives.

1. proboscis
2. oesophagus
3. crop
4. stomach
5. intestine

WHEN DO THEY EAT?

Most species of mosquito are nocturnal: that's why they bother you at night when you're trying to sleep. But there are also some, like the tiger mosquito, which are most active by day.

WHO EATS THEM?

Adult mosquitoes are eaten by swifts, swallows, bats, frogs, lizards and spiders. And mosquito larvae are eaten by lots of other animals!

One female mosquito can store up to three times her own weight in blood in her abdomen!

DRINKING TOOLS

The different diets of male and female mosquitoes mean that their mouths are different too. The most striking feature is the proboscis, the 'straw' they use to slurp up nectar or pierce the skin of the animal whose blood they want to suck. As you might guess, it is longer and more powerful in the blood-drinking females.

SUPER SALIVA

When a mosquito bites you, it squirts saliva on the bite. This keeps its proboscis flexible so it can pierce your skin more easily. The chemicals in its saliva also stop the blood from clotting and prevent you from feeling pain (at least at the time of the bite), so you may not know it's feasting on you.

The octopus is the most intelligent invertebrate in the world. Each one of its tentacles has its own mini-brain!

It also has three hearts and its blood contains a copper-rich substance, which gives it a blueish colour.

COMMON OCTOPUS
Octopus vulgaris

Animal:
INVERTEBRATE

Digestive system:
COMPLETE

Type of diet:
CARNIVORE

Brain food

The digestive system of the octopus is U-shaped and has some very unusual features. First of all, food comes in through the mouth and moves through the oesophagus, which runs through the middle of the octopus's brain! It then reaches the crop, which acts as a storage area. The octopus's stomach can't receive any more food until it has processed the previous 'course' and the waste has been sent to the exit.

Octopuses are excellent hunters, but if they find an animal that's already dead, they won't turn their nose up at it!

While it is not their everyday food, they occasionally practise cannibalism and eat other smaller octopuses. Gulp!

FAVOURITE FOOD

Crustaceans *Molluscs* *Small fish*

WHEN DO THEY EAT?
They are active at night, and spend most of the day hiding in the rocks.

WHO EATS THEM?
Their main predators are large fish and sea birds. Luckily, octopuses have some defence mechanisms. They can camouflage themselves by changing the colour of their skin, and they can also shoot out a jet of ink to confuse an attacker.

1. mouth with beak
2. siphon
3. crop
4. stomach
5. ink sac
6. anus

The liver is the largest organ in the octopus's digestive system. It produces enzymes that help it to digest its prey.

BEAK CAREFUL
Octopus mouths have a hard beak, similar to a parrot's beak. Their jaws are very strong and can bite and tear their prey. The food is then mashed up by the radula, which is like a rough tongue, before moving into the oesophagus.

SIPHON SKILLS
The anus of an octopus is connected to one of its most unusual features: the siphon. This funnel-like tube is used to create a jet of water that propels the octopus through the ocean. What's more, the siphon can squirt out ink if the octopus needs to defend itself against predators, and it's also used to get rid of poo!

Shipworms, from the family Teredinidae, are not worms but molluscs. They are feared by sailors because they love to eat the wood of boats. But some of them eat even stranger things!

In 2019, a new member of the shipworm family was discovered. Its name is *Lithoredo abatanica*, and its diet is even more unusual than wood.

SHIPWORM

Lithoredo abatanica

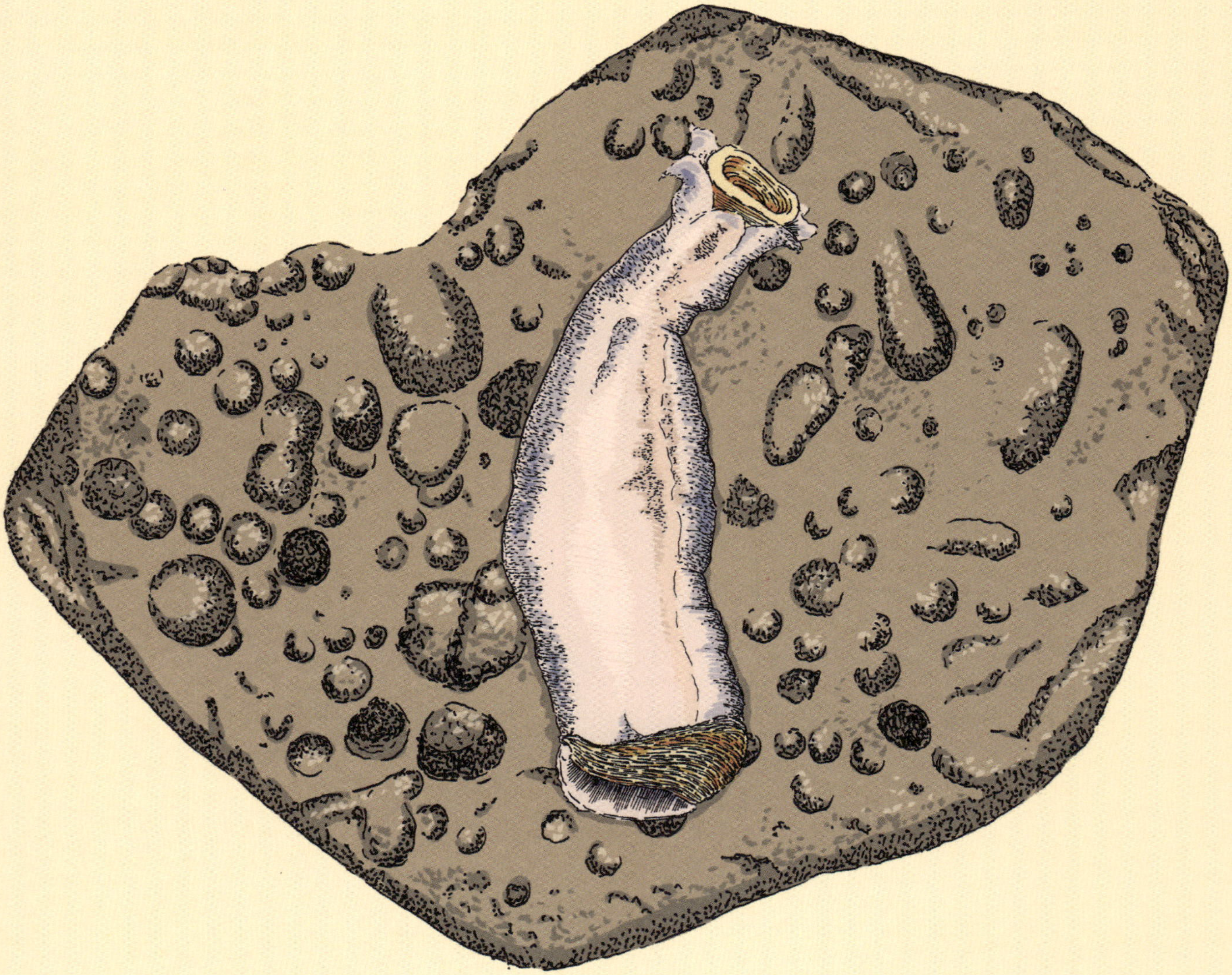

Animal: **INVERTEBRATE** | Digestive system: **COMPLETE** | Type of diet: **LITHOPHAGE**

Let them eat rock!

This species is native to the bed of the Abatan river in the Philippines. It does not dig tunnels by eating wood but eats limestone instead. Yes, you read that right. It eats stone! Obviously, rock doesn't contain any nutrients, so scientists are trying to work out how it's possible for this unusual mollusc to survive on this diet.

FAVOURITE FOOD

Limestone

Some scientists think that the shipworm has bacteria living in its gills, which help it to turn the rock it eats into nutrients.

Another possibility is that they only 'eat' the rock in order to dig tunnels, and they actually feed on plankton, algae or bacteria living in the stone.

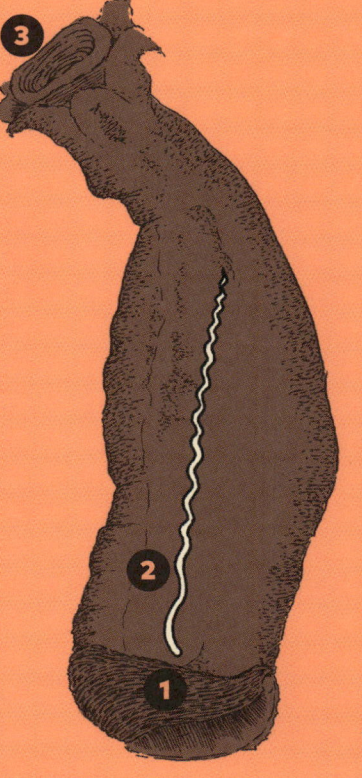

WHEN DO THEY EAT?

Not much is known about the life of *Lithoredo*, but it probably eats constantly. Rocks are hard to swallow!

WHO EATS THEM?

Lithoredo abatanica was only discovered recently, so we don't yet know whether it has predators.

1 mouth
2 digestive organs
3 siphon

Perhaps the London Underground could hire a few of shipworms to dig tunnels?

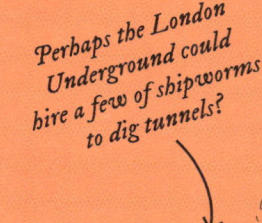

While its wood-eating cousins digest wood in their caecum (the first part of the large intestine), the stone-eating Lithoredo lacks this organ.

DIGGING TUNNELS

Although it is shaped like a worm, the shipworm belongs to the same family as clams and mussels. Its valves (the moving parts of the shell) are located at its front end and are used to dig into rock. They are like jaws with dozens of tiny teeth that can chew through stone.

FROM STONE TO SAND

The rock they eat comes out through the rear of their body as sand. Fine grains are also left in their intestines! If *Lithoredo abatanica* also eats other organisms, such as krill, these grains may be used to grind up its food. A similar thing happens in the gizzard of birds.

It is estimated that these little critters are present in about one out of every ten homes on the planet.

When they come into contact with food, they can contaminate it and cause diseases. They are seen as one of the greatest pests in the animal kingdom!

BLACK COCKROACH
Blatta orientalis

Animal:
INVERTEBRATE

Digestive system:
COMPLETE

Type of diet:
OMNIVORE

Born survivors

People sometimes say that if there was a nuclear war, cockroaches would take over the Earth. It is true that they are much more resistant to radiation than humans, but so are plenty of other insects, such as fruit flies. However, cockroaches are born survivors when it comes to digestion: they can survive for a month without eating! Their secret is in their huge crop. It's the largest organ in their digestive system, and can store food for weeks.

FAVOURITE FOOD

Plant-based waste *Animal-based waste* *Dead insects*

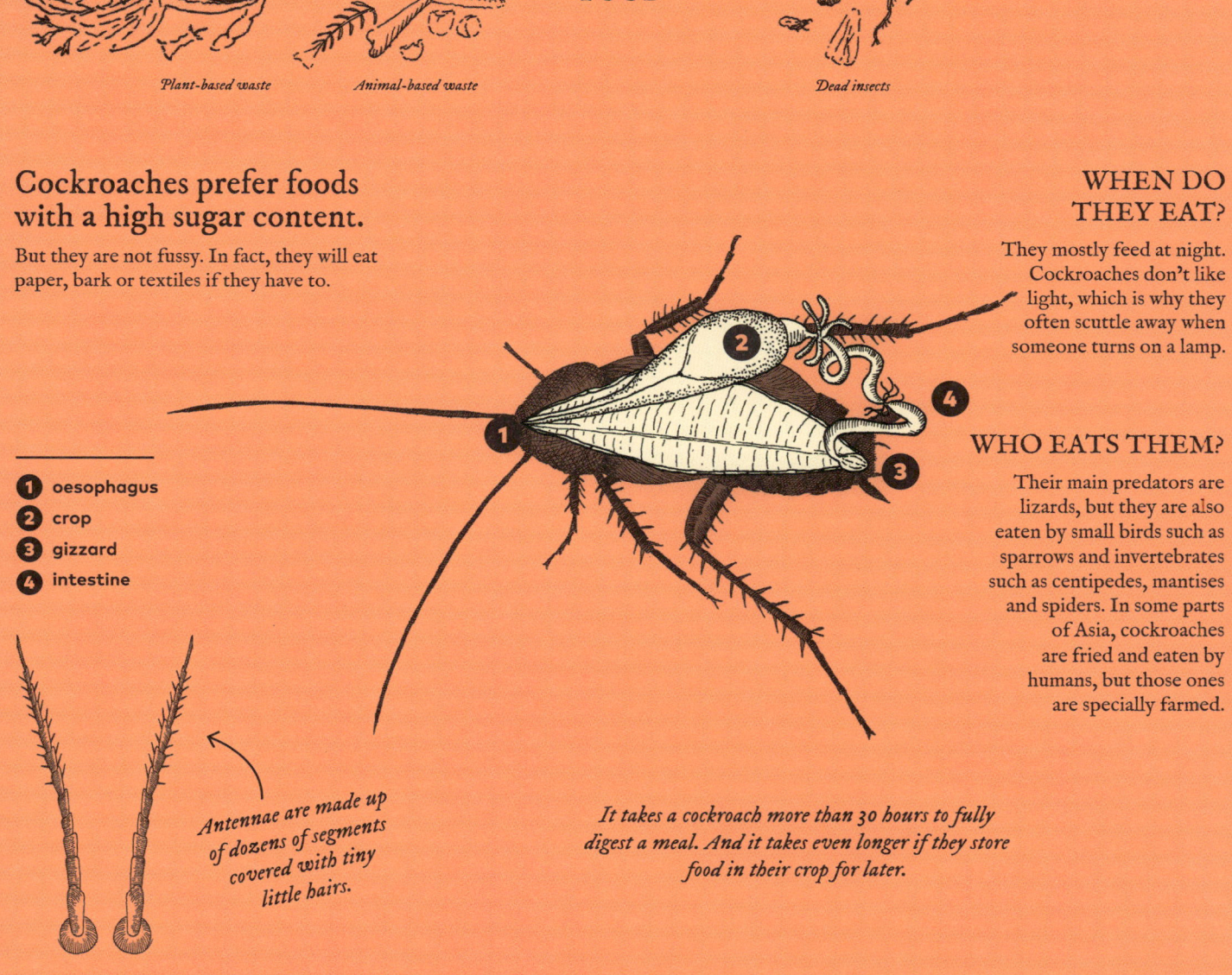

Cockroaches prefer foods with a high sugar content.

But they are not fussy. In fact, they will eat paper, bark or textiles if they have to.

1. oesophagus
2. crop
3. gizzard
4. intestine

WHEN DO THEY EAT?

They mostly feed at night. Cockroaches don't like light, which is why they often scuttle away when someone turns on a lamp.

WHO EATS THEM?

Their main predators are lizards, but they are also eaten by small birds such as sparrows and invertebrates such as centipedes, mantises and spiders. In some parts of Asia, cockroaches are fried and eaten by humans, but those ones are specially farmed.

Antennae are made up of dozens of segments covered with tiny little hairs.

It takes a cockroach more than 30 hours to fully digest a meal. And it takes even longer if they store food in their crop for later.

ALERT ANTENNAE

Cockroaches use their antennae to 'feel' humidity, pressure, temperature, vibration and also to detect smells! This makes them particularly good at searching for tiny crumbs of food.

BACTERIA BESTIES

This species of cockroach (like most others, except for the *Nocticola* genus) never goes anywhere without *Blattabacterium*, a type of bacteria that live in its fatty tissue. These bacteria provide cockroaches with nutrients such as amino acids and vitamins.

Sand dollars are a type of sea urchin with a flattened shell. Most other sea urchins have a rounded shape.

21

Their name comes from their shells (exoskeletons), which are often washed up on beaches. They are shaped rather like coins.

PACIFIC SAND DOLLAR
Dendraster excentricus

Animal:
INVERTEBRATE

Digestive system:
COMPLETE

Type of diet:
OMNIVORE

On the current menu

Pacific sand dollars use undersea currents to bring them food. First, they position themselves vertically in the sand, sinking one side of their body into the ground. Then they point themselves the direction of the current. They use their tiny tentacles (cilia) to catch particles of food from the water and pass them to their mouth, which is in the middle of their body.

  **FAVOURITE FOOD**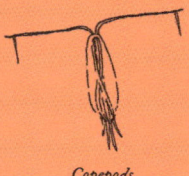

Diatoms   *Crustacean larvae*   *Copepods (small crustaceans)*   *Detritus*

Their favourite foods are so small that they can float in water. Even so, they can take up to 15 minutes to swallow a meal and a couple of days to digest it.

Young sand dollars also swallow sand. They do this to make themselves heavier, so they won't get swept away by the current.

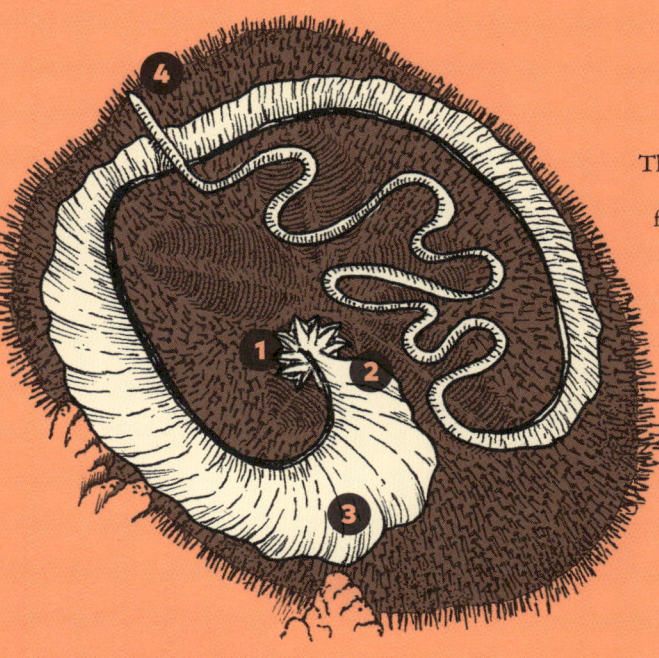

WHEN DO THEY EAT?
They eat at all hours!

WHO EATS THEM?
Their main predators are starfish, crabs, gulls and some fish, such as the starry flounder. Sand dollars can burrow into the sand to hide from danger.

1. mouth
2. oesophagus
3. stomach
4. anus

While the rest of their sea urchin cousins have their anus on the top side of their body, sand dollars have it on the underside.

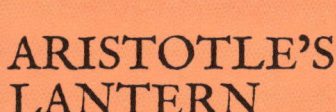

This organ was first described by the Greek philosopher Aristotle. The shape of it reminded him of a lamp.

ARISTOTLE'S LANTERN
This is the name of the sand dollar's mouth parts (most sea urchins have one too). It has five tough 'teeth' made of calcium carbonate, which are used to crush food before it moves into the oesophagus.

TINY HAIRS
The exoskeleton of the sand dollar is covered in spines, but they are shorter, finer and softer than the spines of sea urchins. The spines are covered in tiny little hairs called cilia, which give them a velvety texture and help them move and catch food.

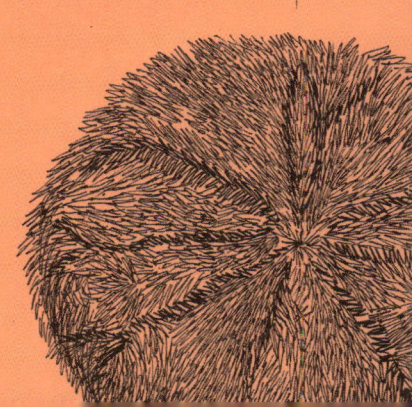

This is the most common species of fly in the world. It lives for about two to four weeks.

22

A fly's eyes can process information seven times faster than human eyes. That's why it's so difficult to hit them with a fly swatter!

HOUSEFLY
Musca domestica

Animal:
INVERTEBRATE

Digestive system:
COMPLETE

Type of diet:
CARNIVORE

A liquid diet

Flies can only take in liquid food. Their mouths cannot deal with particles bigger than 0.045 millimetres. Then how do they eat solid food, you may wonder? It's easy: before swallowing anything, they spit saliva over the food to soften it and break it down. Then dinner is ready!

FAVOURITE FOOD

Rotting organic matter   *Dung*   *Sugar*   *Blood*   *Milk*

Because they spit on their food before eating it, flies can easily contaminate human food and spread diseases.

Fly larvae can also feed on paper and some fabrics, such as wool and cotton!

WHEN DO THEY EAT?

They prefer the drier and warmer hours of the day to go looking for food. When night falls, they find a safe place to rest.

WHO EATS THEM?

There are lots of animals that love to eat houseflies: beetles, spiders, amphibians, reptiles, birds and even other species of flies.

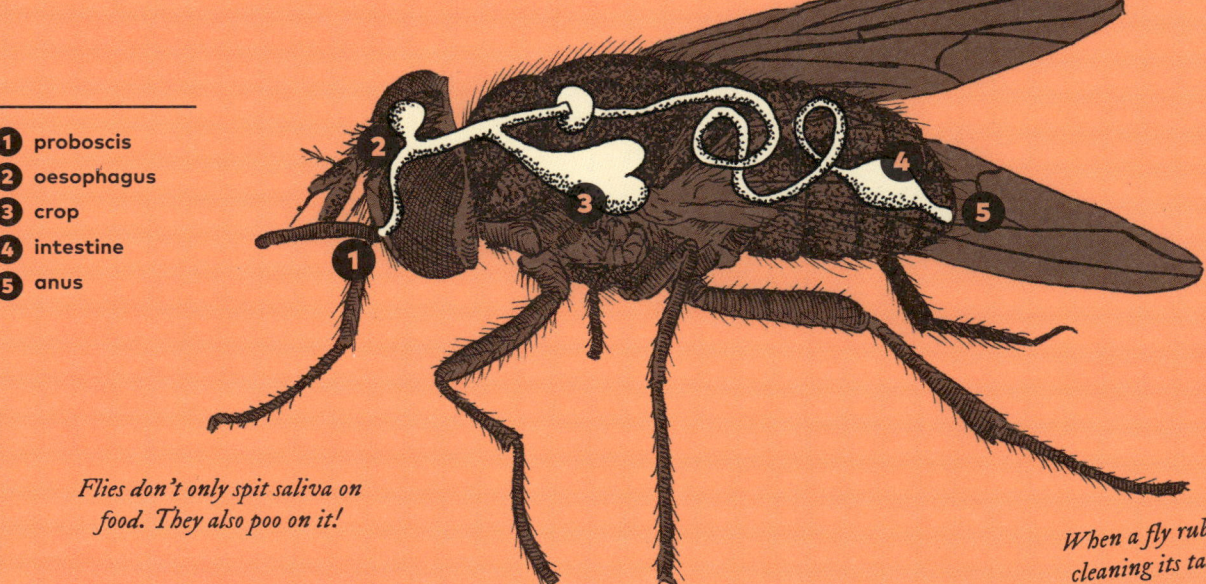

1. proboscis
2. oesophagus
3. crop
4. intestine
5. anus

Flies don't only spit saliva on food. They also poo on it!

When a fly rubs its legs, it's cleaning its taste receptors. Then it's ready to taste the next food it lands on!

A TOOL FOR EATING

The proboscis of a fly is not as long as in other insects like the mosquito, but it does have a very unusual structure. At its tip is the labellum, which is shaped like a tiny sponge covered in grooves called pseudo-tracheae. These soak up the fly's food.

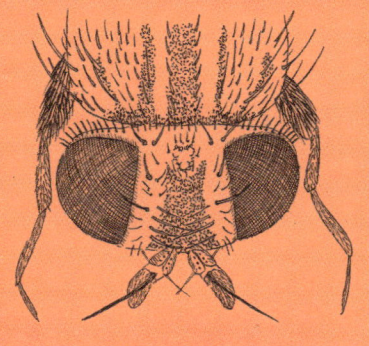

A SENSE OF TASTE

A fly's feet are not only used for walking up walls and across ceilings with their sticky pads. They also contain chemical taste receptors. When a fly lands on food, it can taste whether it is sweet, for example.

Termites are social insects that live in termite mounds. One colony can include thousands of termites.

23

They feed on cellulose, a fibre found in wood. But they can't digit it alone. They need help from another form of life.

TERMITE
Isoptera

Animal:
INVERTEBRATE

Digestive system:
COMPLETE

Type of diet:
HERBIVORE

How to eat wood

Wood is made up of around 50% cellulose, a fibrous substance found in plants. It is tough to digest, but termites have allies to help them. Protozoa, tiny organisms with only a single cell, live in the termite's digestive system and they're the ones who really digest the cellulose. They do this using an enzyme called cellulase. These tiny yet powerful partners help termites to get the nutrients they need.

Most termites eat only one thing: wood, wood and more wood.

Some species specialize in the wood of moist or dry trees, or even wood from human buildings. They can be a real pest, damaging our homes and furniture!

FAVOURITE FOOD

Wood

Rotting plants

WHEN DO THEY EAT?

Termites are temperature-driven. As long as it's nice and warm, they don't stop eating!

WHO EATS THEM?

Other insects such as ants, but also birds, spiders, lizards, amphibians, and even large mammals like anteaters.

Protozoa that help to break down cellulose live in the termite's intestine.

1. mouth parts
2. crop
3. intestine
4. anus

POWER TO THE WORKERS!

The termites in a mound are divided into several 'castes' or groups, which all look different. The workers are the biggest caste. They are in charge of building the mound and feeding on wood. They carry food inside their digestive system and pass it on to other members of the colony.

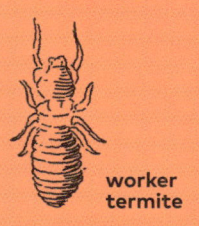
worker termite

soldier termite

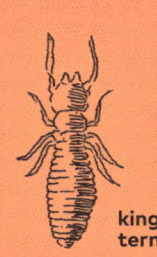
king termite

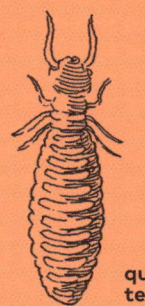
queen termite

PASS IT ON

Trophallaxis is a way of 'passing' food from the mouth of one termite to another. But trophallaxis isn't only used for food. Termites can also get more protozoa by transferring it from anus to mouth — this means eating the poo of another termite!

These mites live in a surprising place: on the faces of humans! Luckily, we can't see or feel them.

They are tiny, only 0.3 to 0.4 millimetres long. And they live for a very short time, no more than two weeks.

FACE MITE

Demodex folliculorum

Animal: **INVERTEBRATE** | Digestive system: **COMPLETE** | Type of diet: **CARNIVORE**

Anywhere there's fresh hair

These mites live inside hair follicles, which are the tiny hollows in our skin where hairs grow out. We have them in our nose, on our forehead and chin, and even in our eyelashes. They live with their heads down inside the follicle, always very close to their source of food.

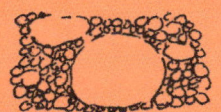
Bodily fluids

FAVOURITE FOOD

Dead skin cells

It seems that what these mites like most is sebum, an oily substance made by hair glands.

They also feed on any dead skin cells they find.

WHEN DO THEY EAT?
By day, they stay in their follicle home where they have all the food they need. By night, they go out looking for a mate.

WHO EATS THEM?
Their lives seem to be pretty quiet. Not even the body of the host they live on (ours!) fights back. And as disgusting as they may sound, these little hitchhikers don't usually cause us any problems.

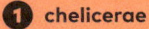

1. chelicerae
2. oesophagus
3. intestine

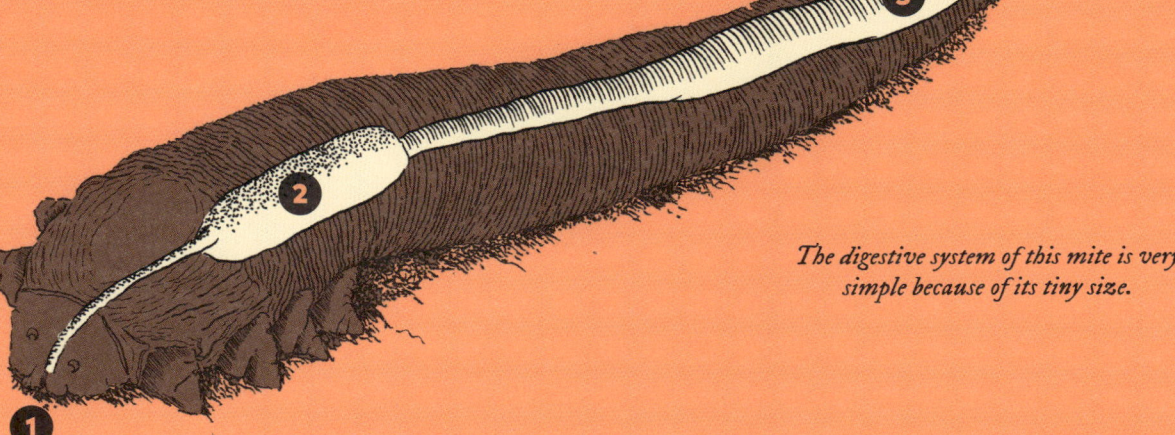

The digestive system of this mite is very simple because of its tiny size.

Not just a pretty face!

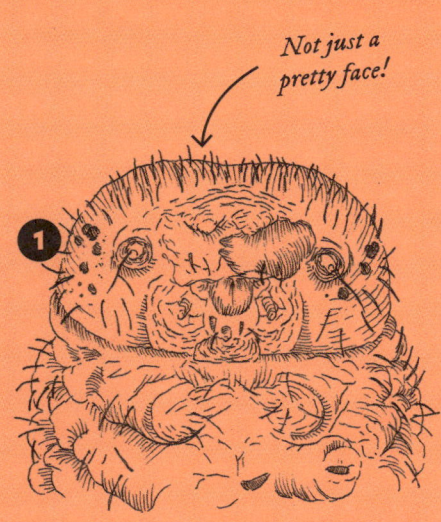

TO POO OR NOT TO POO?

For a long time, it was thought that these mites had incomplete digestive systems, which means they didn't have an anus. Some scientists believed that the waste simply built up inside their digestive system until they died. However, more recent studies appear to show that they do have an anus.

MITE-SIZED MOUTHPARTS

The mouthparts of these mites are called chelicerae: other arthropods such as spiders also have these. They are needle-like in shape and are used to pick up food.

The ancient Greeks and Romans used to eat this species of snail. It's still a popular dish in many countries around the world.

25

Snails are famous for being slow movers. Their top speed is around 0.05 km/h.

ROMAN SNAIL
Helix pomatia

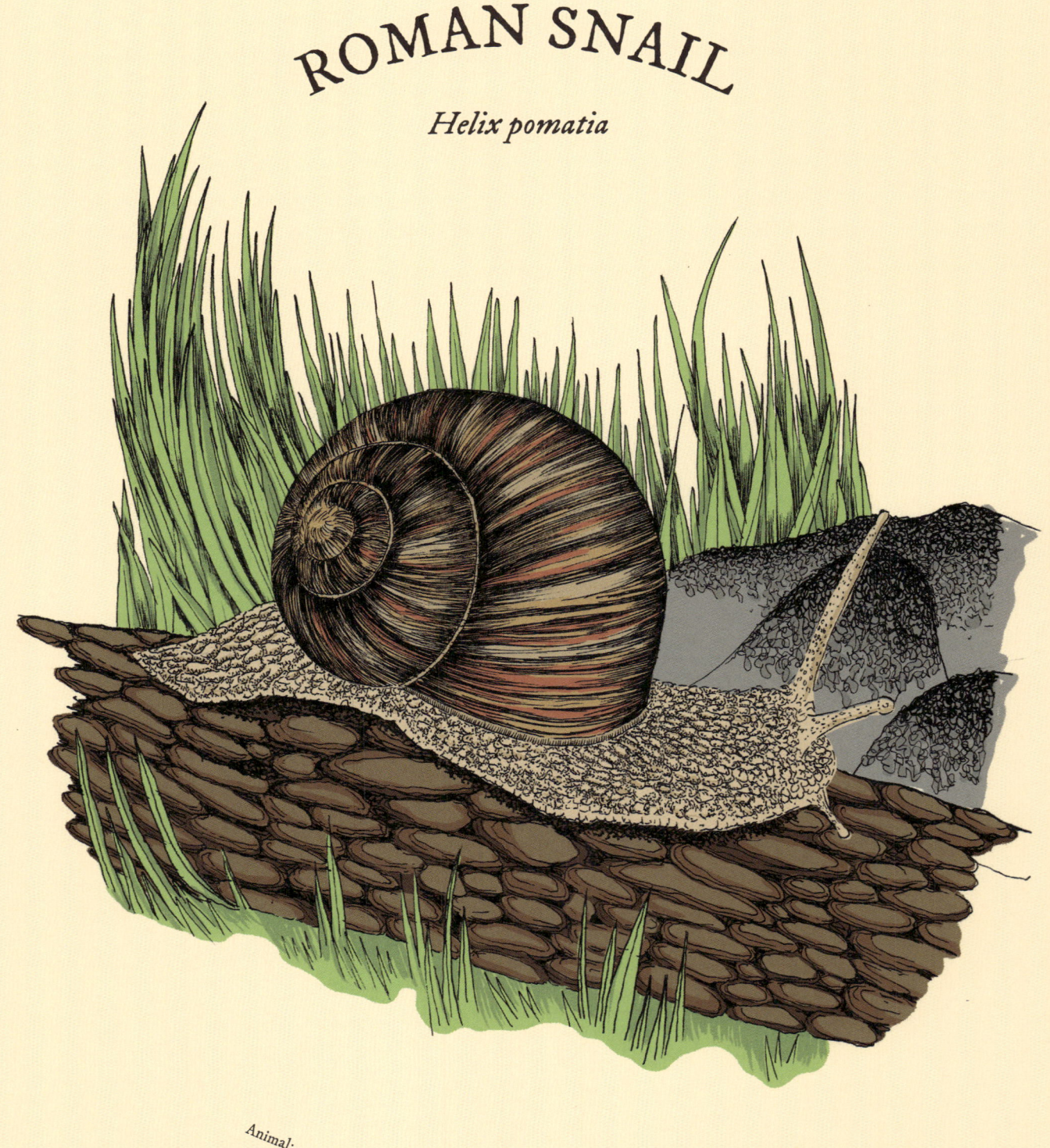

Animal:
INVERTEBRATE

Digestive system:
COMPLETE

Type of diet:
HERBIVORE

Gland of glory

The largest organ in the snail's digestive system is the hepatopancreas, or mid-gut gland. It is rather like a liver and a pancreas joined together, and is more closely involved in digestion than these separate organs are. It produces enzymes that break down food, absorbs and stores nutrients, gathers up waste, and extracts calcium, which the snail needs to make its shell.

FAVOURITE FOOD

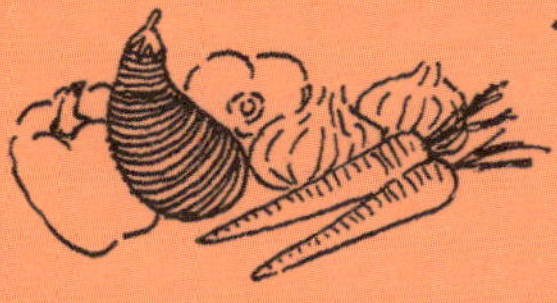

Vegetables  *Leaves*  *Fruit*  *Flowers* 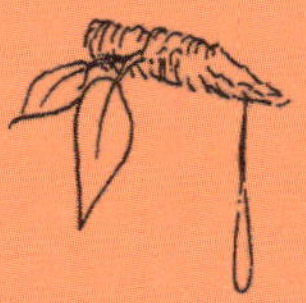 *Sap*

This particular species of snail is especially fond of the green shoots of grapevines.

It is also known as the Burgundy snail, after a region of France with a lot of vineyards.

1. mouth
2. crop
3. hepatopancreas
4. intestine

WHEN DO THEY EAT?
They don't like heat, so they come out looking for food during the cooler hours, especially at night.

WHO EATS THEM?
In addition to humans, they are eaten by birds, toads, rodents, lizards and even other invertebrates such as beetles and centipedes. These prefer young snails or their eggs.

Snail droppings are long and thin. The colour varies, depending on what the snail has eaten.

EATING FOR HOUSE AND HOME

The shell of snails is made of calcium carbonate. This means that their diet must include a lot of this mineral. If they don't get enough, their shell becomes brittle and cracks, which would be life-threatening to the snail. In fact, their home relies on them eating properly.

USEFUL SLIME

The mucus made by snails helps them stick to the ground when moving across slippery surfaces or slopes. But that is not the only function of slime. Among many other things, they use it to return home and to find their way to food sources they've already found, like the trail of pebbles in the story of Hansel and Gretel.

Some species of oyster are considered tasty to eat, while other species, like this one, are prized for the pearls they produce.

Its shell is lined inside with a substance called nacre, or mother-of-pearl, except for the edge, which is black. This gives the oyster its name.

BLACK-LIP PEARL OYSTER

Pinctada margaritifera

Animal:
INVERTEBRATE

Digestive system:
COMPLETE

Type of diet:
PLANKTIVORE

Waving in the water

Oysters have waving hair-like structures on their gills, called cilia. They use these to create a current of water that passes through them, providing the food they need. This process is also good for the ecosystem they live in (coral reefs, in the case of black-lip oysters). This oyster cleans waste products out of the water like the filter in an aquarium!

The oyster catches floating particles of plankton using sticky mucus and then its cilia carry them to its mouth.

After the food is digested, the oyster leaves its droppings on the sea bed.

Oysters produce normal faeces (poo) and also pseudo-faeces, which means particles of grit that their digestive system can't handle. They cover these particles with mucus and push them out of their bodies.

① mouth
② oesophagus
③ stomach
④ anus

FAVOURITE FOOD

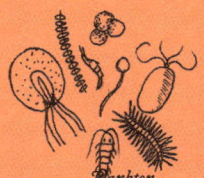

WHEN DO THEY EAT?

Constantly. A single oyster can filter more than 100 litres of water a day.

WHO EATS THEM?

Its shell provides some protection from predators, but the oyster is still eaten by sharks, rays, octopuses, starfish and sea snails.

Contrary to popular belief, pearls are not usually formed around grains of sand.

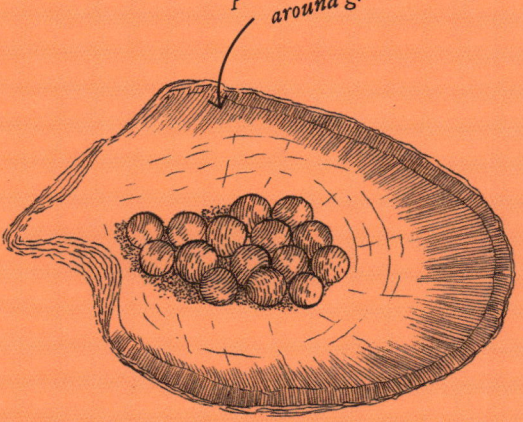

A PEARL OF A TALENT

Oysters have an amazing defence mechanism to protect themselves from irritants like parasites that manage to get inside their shells. When they detect an intruder, they cover it in a substance called nacre (the same one that lines their shells). Layers of nacre build up around the intruder and take on a round shape — this is how pearls are formed. The black-lip oyster is highly prized because it can produce very dark-coloured pearls.

27.
Barn owl

28.
Bee hummingbird

29.
Domestic pigeon

30.
Greater flamingo

31.
Bearded vulture

32.
Bald eagle

33.
Hoatzin

34.
Ostrich

35.
Chicken

VERTEBRATES
Birds

Owls get rid of some of their waste through their mouths. These balls of undigested food are called pellets.

Pellets are formed inside the gizzard, the second chamber of the owl's stomach.

BARN OWL
Tyto alba

Animal:
VERTEBRATE | Digestive system:
BIRD | Type of diet:
CARNIVORE

Down in one!

These ruthless hunters usually swallow their prey whole. Their stomach, which is made up of two sections, separates the soft and edible parts of their prey from the tough and indigestible parts, such as bones, fur, teeth and feathers.

FAVOURITE FOOD

Rats · Mice · Shrews · Rabbits · Small birds

Barn owls feed mostly on small rodents.

A hungry barn owl can eat up to a third of its body weight in a single day. Some farmers use them to control the population of mice and rats in their fields.

1. oesophagus
2. proventriculus (glandular stomach)
3. gizzard
4. small intestine
5. large intestine
6. cloaca

WHEN DO THEY EAT?

They hunt at night. The darkness is not a problem because they rely on their hearing, which is the most powerful in the bird world. The shape of their faces acts like a kind of satellite dish, helping them to detect even the softest sound.

WHO EATS THEM?

They have very few predators. Sometimes they are preyed on by larger owls like the eagle owl. Occasionally they are attacked by other birds of prey, such as eagles, goshawks or sparrowhawks.

The pellets of owls are fairly large, and contain the bones, fur and teeth of the prey they've eaten.

OUT BEFORE IN

Once pellets are formed in the gizzard, they are stored in the proventriculus (the owl's first stomach section). This blocks the owl's digestive tract. It won't be able to eat again until it has regurgitated the pellet, after about ten hours.

PELLET PUZZLE

The pellets of owls and other carnivorous birds are very useful for studying their feeding habits. Breaking apart one of these regurgitated balls is almost like doing archaeology — it reveals the skeleton of the prey.

You could think of these pellets as a rather stinky jigsaw puzzle!

This species of hummingbird from Cuba is the smallest bird in the world. It's barely 5 cm long.

In one day, it can consume half its body weight in food. Of course, that's not much, as it only weighs two to three grams.

BEE HUMMINGBIRD

Mellisuga helenae

Animal: **VERTEBRATE** | Digestive system: **BIRD** | Type of diet: **OMNIVORE**

In-flight feeding

The bee hummingbird, like its other hummingbird cousins, is a highly skilled flyer. It can turn in all directions and hover in the air like a helicopter. This means it can feed on flowers that can't be reached by other animals. Hummingbirds don't need to find a perch — they just hover in front of the flower and eat!

Their main source of food is nectar, the sweet and nutritious juice of flowers.

Nectar is quick and easy to digest (it only takes 15 minutes) so it passes straight into the bird's intestine after they eat it. The hummingbird only uses its stomach if it eats an insect or a spider.

FAVOURITE FOOD

Small insects

Small spiders

Nectar

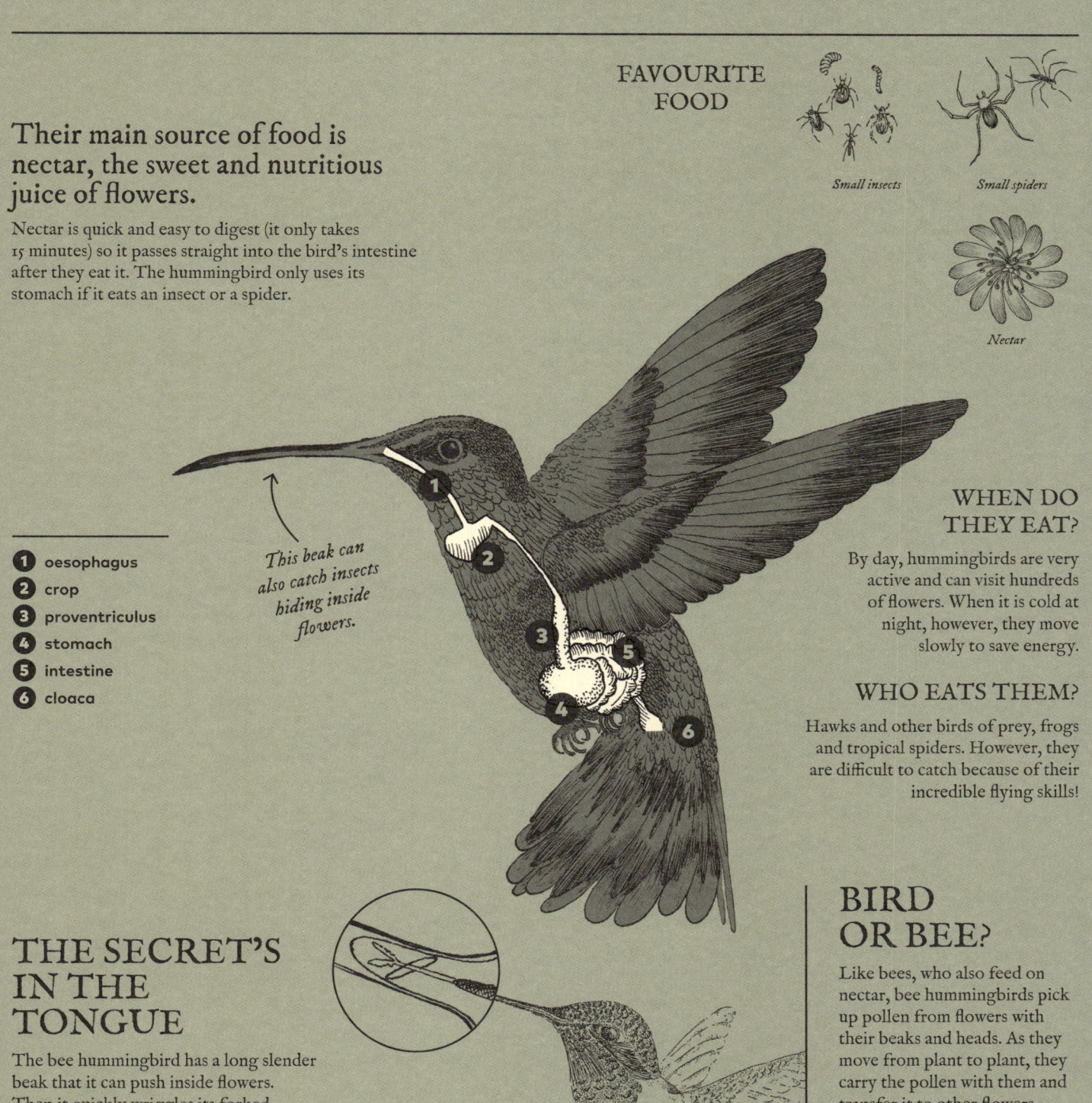

This beak can also catch insects hiding inside flowers.

1. oesophagus
2. crop
3. proventriculus
4. stomach
5. intestine
6. cloaca

WHEN DO THEY EAT?

By day, hummingbirds are very active and can visit hundreds of flowers. When it is cold at night, however, they move slowly to save energy.

WHO EATS THEM?

Hawks and other birds of prey, frogs and tropical spiders. However, they are difficult to catch because of their incredible flying skills!

THE SECRET'S IN THE TONGUE

The bee hummingbird has a long slender beak that it can push inside flowers. Then it quickly wriggles its forked tongue, which is also very long and thin, to reach the tasty nectar.

BIRD OR BEE?

Like bees, who also feed on nectar, bee hummingbirds pick up pollen from flowers with their beaks and heads. As they move from plant to plant, they carry the pollen with them and transfer it to other flowers. They play an important role in plant reproduction.

These birds are very common in big cities all over the world.

29

Female pigeons need a diet with lots of protein and calcium, in order to lay healthy eggs.

DOMESTIC PIGEON
Columba livia

Animal:
VERTEBRATE

Digestive system:
BIRD

Type of diet:
OMNIVORE

Pigeon milk?

Many birds have a crop, which is a cavity that stores food before it goes to their stomach. In the case of pigeons (and a few other birds including turtle doves and flamingos), the crop has another surprising job. It produces a liquid that's high in fat and protein, which the bird uses to feed its chicks. It is a little bit like the milk of mammals, which is why it's called 'crop milk'. It's not milk, however, and both females and males produce it!

FAVOURITE FOOD

Insects   Spiders   Fruit   Grain   Seeds   Leftover human food

In nature, pigeons are mostly herbivores. They eat grains and seeds, plus the occasional small invertebrate.

Feral pigeons in towns and cities feed on any scraps of food they find or are given by humans.

WHEN DO THEY EAT?
They are active during the day, although they tend to avoid the warmer hours. By night, they look for a safe place to rest.

WHO EATS THEM?
Birds of prey are their main predators, including hawks, sparrowhawks, owls and eagles. They are also hunted by mammals such as cats and raccoons.

Pigeon droppings are very acidic. In cities they can cause damage to buildings and statues.

1. oesophagus
2. crop
3. stomach
4. gizzard
5. intestine
6. cloaca

A UNIQUE WAY TO DRINK

Pigeons drink by sucking up water with their heads down. They can swallow it in that position without needing to lift their heads up. You might not think that's a big deal, but pigeons and doves are the only birds that can do this!

When it's hot, pigeons need water to cool down their bodies.

RATS WITH WINGS?

Humans often dislike pigeons because it's said that their droppings and feathers can spread diseases, and they can also carry parasites. These things are true, but they are minor risks and their bad reputation is rather unfair.

Scientists think that the classic flamingo pose, standing upright on one leg, is a way to keep their body temperature steady.

30

Flamingos are very social. A single flock can contain thousands of birds.

GREATER FLAMINGO

Phoenicopterus roseus

Animal: **VERTEBRATE** | Digestive system: **BIRD** | Type of diet: **OMNIVORE**

Big and beaky

Unlike the vast majority of birds, a flamingo has a beak that's larger at the bottom and smaller at the top. It may look as if it's been designed the wrong way around, but there's a good reason for it. Flamingos eat with their heads upside down: they stretch their necks down into the water, and stir up the bottom so any pieces of food float around. Then they eat!

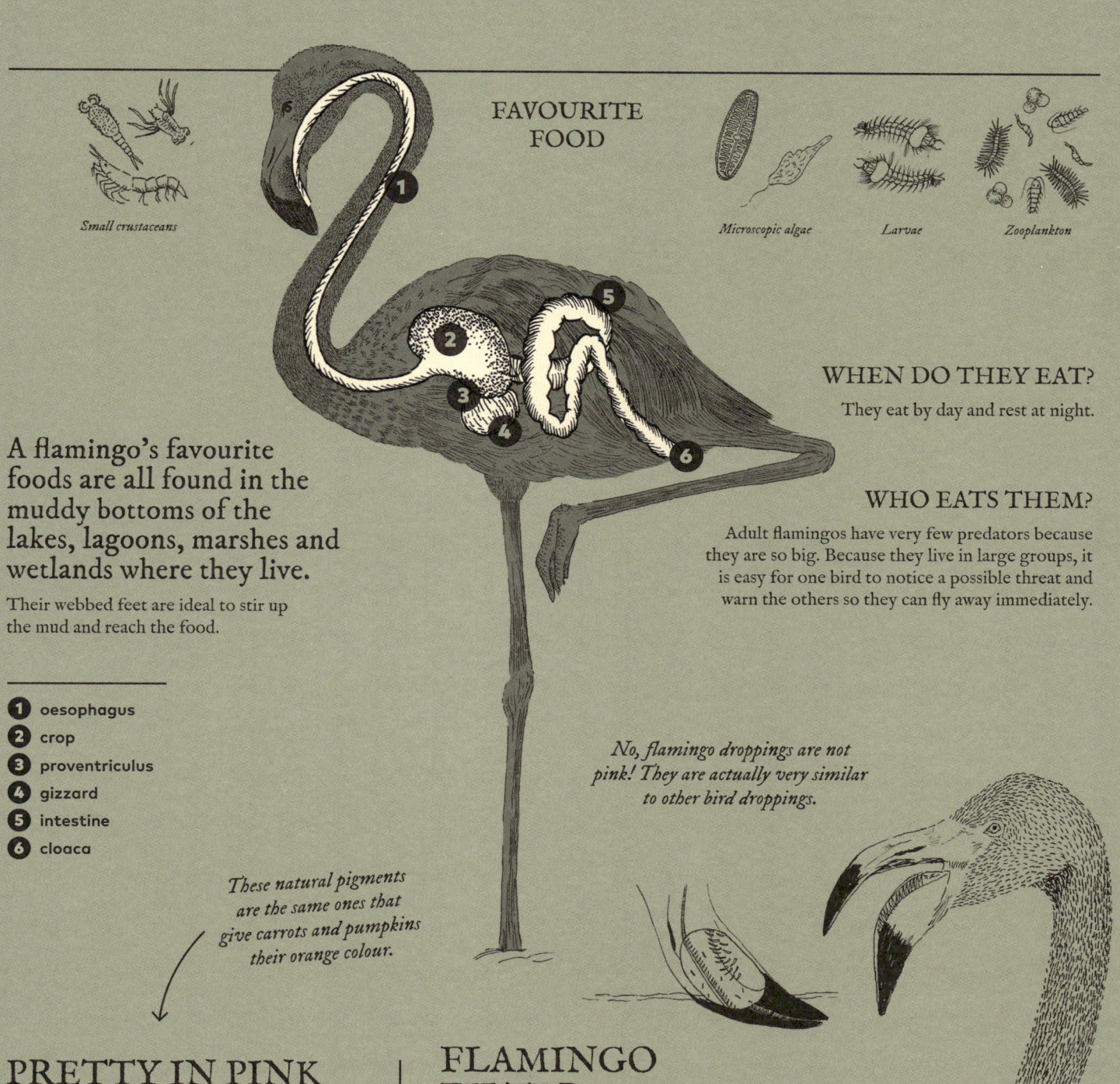

FAVOURITE FOOD

Small crustaceans · *Microscopic algae* · *Larvae* · *Zooplankton*

WHEN DO THEY EAT?
They eat by day and rest at night.

WHO EATS THEM?
Adult flamingos have very few predators because they are so big. Because they live in large groups, it is easy for one bird to notice a possible threat and warn the others so they can fly away immediately.

A flamingo's favourite foods are all found in the muddy bottoms of the lakes, lagoons, marshes and wetlands where they live.

Their webbed feet are ideal to stir up the mud and reach the food.

1. oesophagus
2. crop
3. proventriculus
4. gizzard
5. intestine
6. cloaca

No, flamingo droppings are not pink! They are actually very similar to other bird droppings.

These natural pigments are the same ones that give carrots and pumpkins their orange colour.

PRETTY IN PINK
Young flamingos have whitish feathers. As they grow up, their feathers turn pink because of pigments called carotenoids, which come from the algae and crustaceans that they eat.

FLAMINGO FILTER
Flamingos fill their beaks with water and mud when they eat. Inside their beaks, they have rows of bony plates called lamellae. These act like a filter, allowing the bird to sieve out tasty larvae and crustaceans to eat, while getting rid of smaller particles such as grains of sand.

This species of vulture often throws bones from a height to break them and feed on the marrow inside.

They live in rugged mountain regions, where it is easy for them to feed in this way.

BEARDED VULTURE
Gypaetus barbatus

Animal:
VERTEBRATE

Digestive system:
BIRD

Type of diet:
OSTEOPHAGE

Bone-breakers

After apex predators have eaten their fill of prey such as goats or antelope, and any scavengers have stripped the carcass to the bone, it is the turn of the bearded vultures, also known as lammergeiers. They can swallow several kilos of bones, but if the bones are too big, they break them up by throwing them on rocky ground to smash them into smaller pieces. The powerful acid in their stomach helps them to digest bones within 24 hours.

FAVOURITE FOOD

Bones | Fur | Carrion meat  | Tortoises | Mice | Small reptiles

Bones make up about 80% of the diet of these birds but they eat other foods too.

For example, young vultures cannot digest bones very well, so they eat more fur and carrion meat instead.

1. oesophagus
2. crop
3. proventriculus
4. gizzard
5. small intestine
6. large intestine
7. cloaca

WHEN DO THEY EAT?

They fly over the mountains during the day, looking for bones to feed on.

WHO EATS THEM?

These powerful birds have a wingspan of nearly three metres and weigh up to seven kilos. They do not have any natural predators. Their biggest threat is humans and the things we create, such as electric fences.

Their droppings are white and dry and have a high mineral content. They look a bit like chalk!

WATCH OUT FOR FALLING TORTOISES

Bearded vultures like to feast on a tasty tortoise when they get the chance. To break open its shell, they drop it from a great height, just like they do with bones. According to legend, the ancient Greek playwright Aeschylus was killed by a tortoise dropped by a bearded vulture.

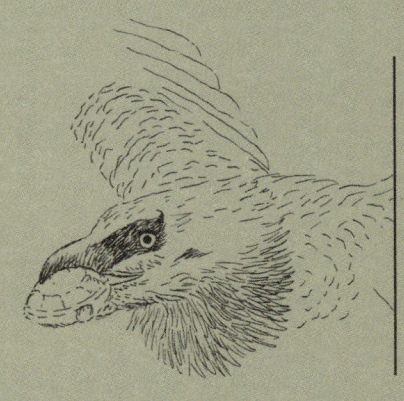

THE BALD TRUTH

Most vulture species push their heads and necks inside rotting carcasses to reach the meat, which is why they don't have feathers on those parts of their body. But bearded vultures don't need to do this because they mostly eat bones, so they can grow a beautiful mane of feathers.

This majestic bird is the national symbol of the United States. It features on the American seal.

Although its head is covered in feathers, it's called the bald eagle because the word 'bald' in old English meant 'white', not 'hairless'.

BALD EAGLE

Haliaeetus leucocephalus

Animal:
VERTEBRATE

Digestive system:
BIRD

Type of diet:
CARNIVORE

Someone else's dinner

Bald eagles are known for their ability to save energy when hunting for food. Often they are very sneaky about this. They let another creature, such as an osprey or fox, do the dirty work and hunt for a tasty fish or rabbit while they watch. Then they swoop down on the hunter and steal the prey, or attack them until they drop what they've caught. Cheeky thieves!

FAVOURITE FOOD

Trout · Salmon · Catfish · Carrion · Eels · Water birds · Small mammals

When bald eagles hunt for their own prey, they prefer to catch fish that swim close to the surface or leap above the water, such as salmon and trout.

However, they are rather lazy, so they will eat carrion if it is available. A free meal is always welcome!

An eagle's beak is made of keratin, like human fingernails. It keeps growing throughout the bird's life.

They can survive several days without eating, thanks to the food they store in their crop.

WHEN DO THEY EAT?

They are most active during the day. They spend most of their time (90%) resting, but they will eat whenever they get a chance.

WHO EATS THEM?

Adult bald eagles have no natural predators. But their eggs and chicks can be targets for crows, magpies, foxes and wolves.

1. oesophagus
2. crop
3. proventriculus
4. gizzard
5. small intestine
6. large intestine
7. cloaca

BEAK AND CLAWS

The bald eagle's beak, curved and sharp as a hook, looks quite fearsome but it's not used for hunting. Hunting is done with the deadly claws on its feet. The eagle uses its beak to rip its meat into pieces.

EAGLE OR TURKEY?

Benjamin Franklin, one of the founding fathers of the USA, did not want the bald eagle to become a symbol of the country because of its reputation for stealing. His suggestion was the turkey!

This tropical bird from South America is known by many names, including *chenchena* (in the Llanos region), *pava serere* (in Bolivia), *shansho* (in Peru), *guacharaca de agua* (in Venezuela) and 'stinky turkey' (in Colombia)!

The hoatzin's spiky fan-shaped crest gives it a rather messy look — a little like bed hair!

HOATZIN
Opisthocomus hoazin

Animal:
VERTEBRATE | Digestive system: **BIRD** | Type of diet: **HERBIVORE**

Can a bird be a ruminant?

The hoatzin has a unique digestive system for a bird. Most of the digestion process happens in its crop and lower oesophagus, which are huge compared with those of other birds. They contain special bacteria that break down and ferment the plants that the hoatzin eats. This is similar to what happens in the stomachs of cows and other ruminants. The hoatzin's stomach and gizzard, on the other hand, are much smaller than in other birds.

FAVOURITE FOOD

Leaves *Fruit* *Flowers*

More than 80% of the hoatzin's diet consists of leaves, usually from plants such as arums and mangroves.

Every so often, they swallow an insect along with the plants they eat, but not on purpose!

WHEN DO THEY EAT?

With their spiky crests, they may look ready for a night out, but in fact they're most active during the day.

WHO EATS THEM?

Birds of prey such as the black hawk, along with snakes and monkeys.

1. upper oesophagus
2. crop
3. lower oesophagus
4. proventriculus
5. gizzard
6. intestine and cloaca

Bacteria from the bird's crop are passed from parent to child. But how? Well, the parent bird vomits up a sticky substance full of bacteria and feeds it to its babies. Yuck!

Straight after eating, the crop and oesophagus of a hoatzin make up 25% of its body weight.

FERMENTING, NOT FLYING

The unique digestive system of the hoatzin also has disadvantages. The top part of its digestive tract is so big that there's not much space for its sternum (breastbone) or pectoral (chest) muscles, which birds use for flying. And when the hoatzin eats, all the food settles in its crop, pushing its centre of gravity too far forwards. For these reasons, the hoatzin can't fly, although it can glide for short distances.

A BIRD WITH BAD BREATH

As a side effect of the fermenting food in its digestive system, the hoatzin has very smelly breath, which reeks of manure! This is why it's called a stinky turkey in Colombia.

The wings of the ostrich are not used for flying but for display and to help them balance while running. An ostrich can reach speeds of 70 km/h!

34

It is the tallest and heaviest bird in the world, with the biggest ones measuring up to three metres in height and weighing almost 200 kg.

OSTRICH
Struthio camelus

Animal:
VERTEBRATE

Digestive system:
BIRD

Type of diet:
HERBIVORE

The camel of the bird world

The scientific name for the ostrich includes the term *camelus*, which means camel. Ostriches have a lot in common with camels, including long necks, living in dry places, and an ability to go without water for a long time. Their digestive system is specially adapted to process the fibre from the plants they eat – the same is true of pseudo-ruminants such as camels. Unlike most birds, the ostrich has no crop. It ferments the fibrous part of its food in its large intestine, which is extremely long, measuring about 16 metres.

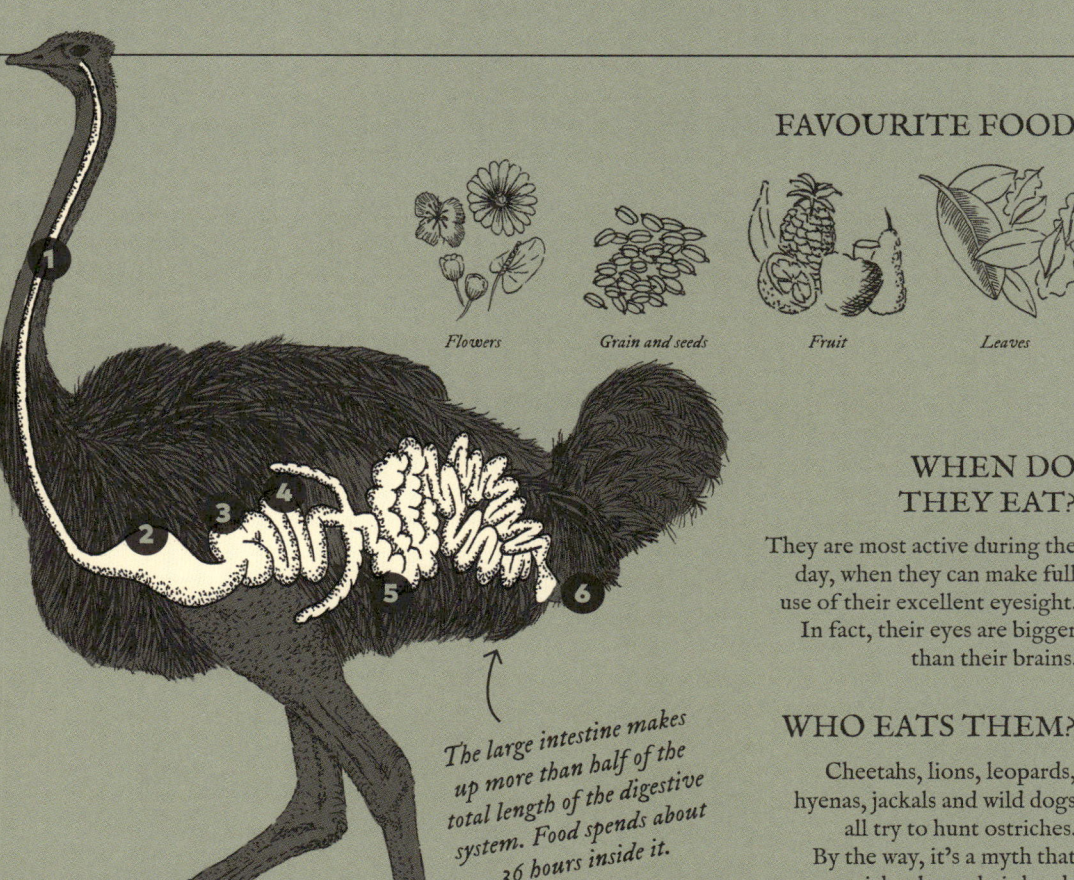

FAVOURITE FOOD

Flowers   *Grain and seeds*   *Fruit*   *Leaves*

An adult ostrich eats 2 to 4 kg of plants a day.

It can also eat insects, carrion and even snakes and small rodents, which means it is sometimes considered an omnivore.

1. oesophagus
2. proventriculus
3. gizzard
4. small intestine
5. large intestine
6. cloaca

The large intestine makes up more than half of the total length of the digestive system. Food spends about 36 hours inside it.

WHEN DO THEY EAT?

They are most active during the day, when they can make full use of their excellent eyesight. In fact, their eyes are bigger than their brains.

WHO EATS THEM?

Cheetahs, lions, leopards, hyenas, jackals and wild dogs all try to hunt ostriches. By the way, it's a myth that ostriches bury their heads in the ground when there's danger. Instead, they defend themselves with kicks that can be deadly!

You might say this meal was heavy-going!

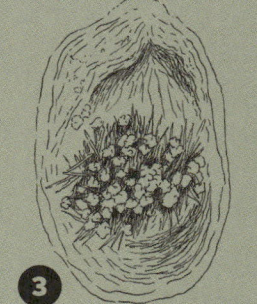

SAND AND STONES

When feeding, ostriches also eat sand and stones. Like some other birds, they store these in their ventricle or gizzard and use them to crush up food. Because they're so big, ostriches can carry around 1.5 kg of stones at once.

AN ORDERLY SPECIES

Most birds use their cloaca to get rid of solid and liquid waste at the same time. The ostrich does this too, but in an orderly fashion: first the pee, then the poo!

Male chickens are called cockerels or roosters. The females are called hens and the young are chicks.

Humans began to domesticate chickens thousands of years ago. The *Gallus gallus domesticus* subspecies is now the world's most common bird.

CHICKEN
Gallus gallus domesticus

Animal: **VERTEBRATE** | Digestive system: **BIRD** | Type of diet: **OMNIVORE**

No longer needed

Many of the features of a typical bird digestive system have disappeared in domesticated chickens. For example, they are usually given crushed feed to eat, which means they don't need stones or grit in their gizzards. Also, wild chickens use their crop to store food in case they can't find any later, but domestic birds get a steady flow of food.

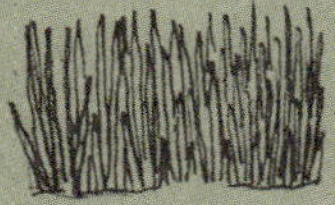

FAVOURITE FOOD

Chicken feed

Grass

Grain

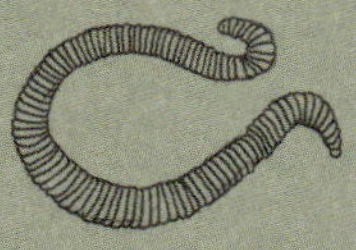

Small invertebrates

Chickens feed on grains and pulses. They also need calcium to form the shells of their eggs

They can also eat invertebrates such as worms and insects, and even small reptiles and rodents.

Chickens cannot taste sweet things. But they can taste salt and they don't usually like it!

WHEN DO THEY EAT?

Chickens are diurnal. By night, they go back to their hen house and sleep for about eight hours, just like us.

WHO EATS THEM?

We humans eat chickens and their eggs. Foxes and other predators will also eat them if they get a chance.

1. oesophagus
2. crop
3. proventriculus
4. gizzard
5. small intestine
6. large intestine
7. cloaca

A CROP FOR ALL SEASONS

While it is true that domestic chickens no longer need their crop for storage, the crop also does other jobs. It softens food, which is particularly useful with grain as it can stay in the crop for up to 12 hours before continuing through the chicken's system. The crop also has a protective function: it is filled with 'good' bacteria that fight against any 'bad' bacteria, viruses or fungi that might cause infections.

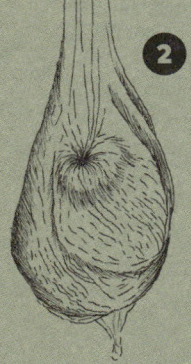

This is much more than an emergency snack box!

VERTEBRATES

Monogastric

The human digestive system is very versatile. It can easily digest both plants and meat.

36

The longest part of our digestive tract is the small intestine. In an adult, it's about 7 metres long and 3 cm across. Luckily it's all neatly rolled up!

HUMAN

Homo sapiens sapiens

Animal: **VERTEBRATE** | Digestive system: **MONOGASTRIC** | Type of diet: **OMNIVORE**

Now we're cooking!

The use of fire to cook food was a vital stage in the evolution of human beings. Among many other things, it led to big changes in our digestive system. Because cooked foods are easier to eat and digest, we evolved to have less powerful jaws and teeth and a shorter intestinal tract, in comparison with other apes.

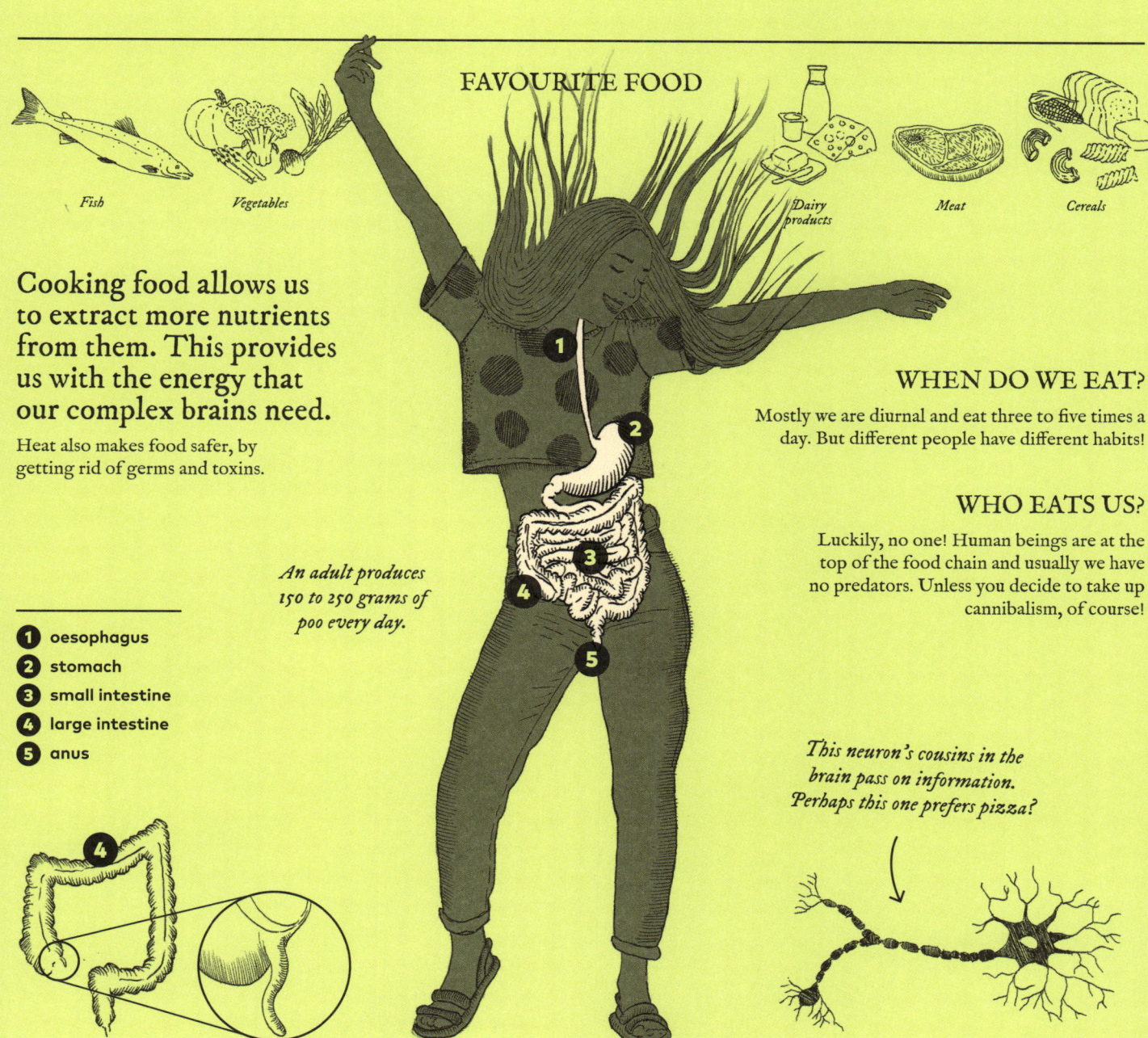

FAVOURITE FOOD: Fish, Vegetables, Dairy products, Meat, Cereals

Cooking food allows us to extract more nutrients from them. This provides us with the energy that our complex brains need.

Heat also makes food safer, by getting rid of germs and toxins.

An adult produces 150 to 250 grams of poo every day.

1. oesophagus
2. stomach
3. small intestine
4. large intestine
5. anus

WHEN DO WE EAT?

Mostly we are diurnal and eat three to five times a day. But different people have different habits!

WHO EATS US?

Luckily, no one! Human beings are at the top of the food chain and usually we have no predators. Unless you decide to take up cannibalism, of course!

This neuron's cousins in the brain pass on information. Perhaps this one prefers pizza?

A USELESS ORGAN?

The appendix, at the entrance to the large intestine, is a small organ whose job in the digestive system is still unclear. For a long time, people believed it didn't do anything, although it is now known that it helps to store gut bacteria. However, if the appendix gets infected, it must be removed immediately with an operation.

A BRAIN IN THE BELLY?

The human digestive system contains a network of more than one hundred million neurons (cells that transmit information) that functions independently. Outside of the brain, no other part of the human body has as many neurons. In fact, there are more neurons in our digestive system than in a rabbit's brain!

Kangaroos are marsupials, a group of mammals whose babies finish growing inside a pouch attached to their mother's body.

37

The red kangaroo is the largest mammal in Australia and the largest marsupial in the world. It can reach 1.8 metres in height and weighs around 90 kg.

RED KANGAROO
Macropus rufus

Animal: **VERTEBRATE** | Digestive system: **MONOGASTRIC** | Type of diet: **HERBIVORE**

Ruminant or rumin-not?

Kangaroos sometimes regurgitate the food they've eaten to chew it as cud and then swallow it again. However, this isn't true rumination as they only do it occasionally and it's not a vital part of their digestive process. This habit is called merycism. Also, unlike ruminants, kangaroos are monogastric.

FAVOURITE FOOD

Leafy plants 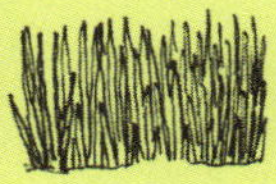 *Grass*  *Shrubs*

WHEN DO THEY EAT?
They prefer to be active in the evening or at night to avoid the heat. During the hotter hours of the day, they laze in the shade of trees and bushes.

WHO EATS THEM?
Adult kangaroos are so big, strong and fast that they are not easy prey for any animal, not even dingoes, the largest predators in Australia. Human hunters are their main threat.

The diet of kangaroos changes with the seasons.
During the rainy season, they feed on leafy plants, then settle for shrubs and grass during the dry season.

1. oesophagus
2. forestomach (sacciform chamber)
3. hindstomach (tubiform chamber)
4. intestine
5. anus

Kangaroo poo is very dry because they live in dry places and must extract as much water as possible from what they eat.

A SUPER STOMACH
A kangaroo, like a ruminant, eats a lot of high-fibre plants. The secret to digesting these plants lies in its stomach, which is larger and longer than other monogastric stomachs. In one of its two chambers, bacteria and other microorganisms ferment the fibre, while the other chamber finishes the job with acids and enzymes. Kangaroos extract 70% of their energy from fibre, more than true ruminants do!

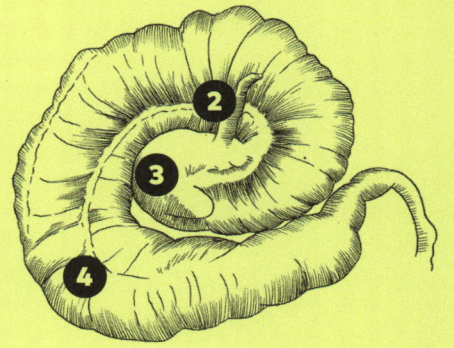

ECO-FRIENDLY FARTING!
Kangaroo digestion does not create methane, a polluting gas released by cattle and other ruminants. In fact, scientists are studying the possibility of using the digestive bacteria of kangaroos in cattle to reduce their methane output.

It is estimated that there are more than 600 million cats in the world. After dogs, they are the world's most popular pets.

38

Cats need to preserve their energy with long naps, and can snooze for up to 20 hours a day.

DOMESTIC CAT
Felis silvestris catus

Animal: **VERTEBRATE** | Digestive system: **MONOGASTRIC** | Type of diet: **CARNIVORE**

Carnivores with a side of veggies

Cats are talented hunters and the most fearsome on the planet in proportion to their size. To catch their prey, they have a powerful sense of smell (thirty times better than humans), excellent night vision, outstanding hearing and great agility (they can leap up to five times their height). But they also nibble on herbs or plants, which help their digestion.

FAVOURITE FOOD

Fish · Dry food · Birds · Tinned food · Rodents

Throughout history, cats have been used to get rid of mice and rats. This is one of the reasons why they have become such popular pets.

However, when they live with humans, they rely on the food we give them. Being cute is also a great advantage!

1. oesophagus
2. stomach
3. small intestine
4. large intestine
5. anus

WHEN DO THEY EAT?

Domestic cats eat at any time, but in the wild, they tend to be nocturnal. That's why they often start getting playful at night.

WHO EATS THEM?

If they live outdoors, they can fall prey to larger predators, depending on where they live. These include coyotes, foxes, crocodiles and snakes.

Cats instinctively poo in sand or soil. This hides their scent from possible predators or rival cats.

Although a cat's tongue has many talents, it can't taste sweet things.

A FLICK OF THE TONGUE

A cat's tongue is quite unusual. If you have ever been licked by a cat, you will have noticed that it's quite raspy. It is covered in tiny spines made of keratin, the substance that our nails and hair are made of. These help the cat to tear through the hair, feathers or fur of its prey.

HAIR TODAY?

Their rough tongue is also used for grooming. When cats lick themselves, they get clean and remove dead hair at the same time. As they do this, they often swallow some hair accidentally. They can't digest it, so they throw it up as a hairball.

This creature is responsible for one of the most prized coffees of the world, kopi luwak or civet coffee. You may be surprised when you hear how it is made!

39

It also produces a substance called civet musk, which is used to make perfumes. Fortunately this is now quite rare, because it's extracted in a way that is cruel to the civet.

ASIAN PALM CIVET

Paradoxurus hermaphroditus

Animal:
VERTEBRATE

Digestive system:
MONOGASTRIC

Type of diet:
OMNIVORE

104

Don't ask where that coffee comes from!

Civets are rather like a cross between a cat, a ferret and a raccoon, and are famous for eating ripe coffee berries. However, the coffee bean inside the berry is not digested and comes out whole in the civet's poo. The stomach juices and enzymes of the civet change the taste of these beans. The beans are collected and used to make an expensive type of coffee with a very distinctive taste and smell.

FAVOURITE FOOD

Mango

Figs

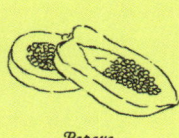
Papaya

Sugar palm flower

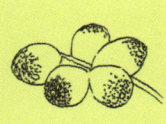
Coffee berries

Small land animals

Civets are omnivores but their favourite food is fruit.

They are excellent climbers and can reach the best fruit from their favourite trees. But usually they will eat whatever is available, including worms and seeds.

WHEN DO THEY EAT?

They go out looking for food in the darkness of the night, so they can avoid predators. They spend the day hiding in trees or among rocks.

WHO EATS THEM?

Their predators include tigers, leopards, snakes and crocodiles. Human beings also breed them in captivity.

1. oesophagus
2. stomach
3. small intestine
4. large intestine
5. anus

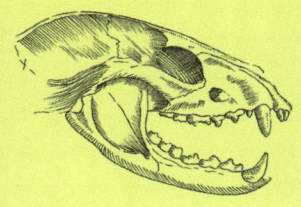

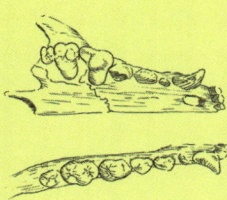

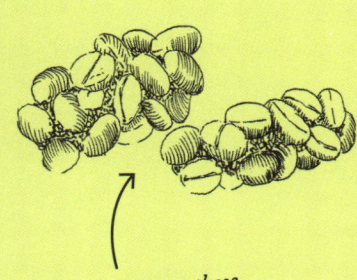

Coffee made from these digested beans is the most expensive in the world.

A LITTLE BITE

The teeth of the palm civet are adapted to its plant-heavy diet, more so than the other species of civet. For example, its carnassial teeth, which carnivores use to rip up meat, are quite small, and the rest of its teeth are not very strong or sharp.

A CRUEL BUSINESS

Unfortunately, the demand for civet coffee has led to these animals being farmed and exploited. The civets are forced to live in crowded cages and are only fed on coffee berries, which can lead to malnutrition. Many animal welfare charities want to stop this abuse.

This species has been living on Earth for at least sixteen million years. It's practically a living fossil!

40

The great white shark can reach 8 metres in length and 2 tonnes in weight. Along with the orca, it is the largest predator in the ocean.

GREAT WHITE SHARK
Carcharodon carcharias

Animal: **VERTEBRATE** | Digestive system: **MONOGASTRIC** | Type of diet: **CARNIVORE**

An empty stomach

Great white sharks are such keen eaters that they sometimes swallow things they can't digest. Turtle shells, tyres and cameras are some of the things that have been found in their stomachs. But if they eat something they shouldn't have, they have a special skill called stomach eversion. Basically, it means pushing their stomach out through their mouth to empty it, rather like turning a bag inside out.

FAVOURITE FOOD

Seals · Sea lions · Elephant seals · Dolphins · Tuna · Whale carcasses

Sharks usually attack their prey from below, swimming upwards at full speed until they get a good bite.

Despite their scary reputation, in real life there are only a small number of shark attacks on humans every year. We are definitely not their favourite prey!

WHEN DO THEY EAT?

They usually hunt and feed by day. But they don't spend their nights dozing in a bed of seaweed. They need to swim constantly to be able to breathe.

WHO EATS THEM?

The great white shark is an apex predator, which is an animal at the top of the food chain. Its only threat is from human beings.

1. oesophagus
2. stomach
3. intestine (spiral valve)
4. anus

The spiral valve slows down the digestion process. This means the shark gets more nutrients without needing extra-long intestines.

Sharks do not have teeth for chewing. They only need them for tearing up meat before they swallow it.

ARMED TO THE TEETH

The jaws of a great white are as terrifying as they look. Its bite is enormously strong (twenty times stronger than our own), but the damage is caused mostly by its sharp triangular teeth, each around 7 cm long. They are grouped together in multiple rows. Sharks often lose teeth but new ones soon grow to replace them.

Remoras spend most of their lives hitching a lift on larger sea creatures, such as sharks and turtles.

41

When they find a new host and stick themselves to it, they can stay with it for several months.

COMMON REMORA
Remora remora

Animal: **VERTEBRATE** | Digestive system: **MONOGASTRIC** | Type of diet: **CARNIVORE**

A well-matched pair

Symbiosis is a type of relationship between two different species, in which neither species is harmed and at least one benefits. That is exactly what happens with remoras and sharks. The remora's suction cup allows it to attach itself to the shark, which provides it with transport and protection, and also helps it find food. It's a surprisingly easy life!

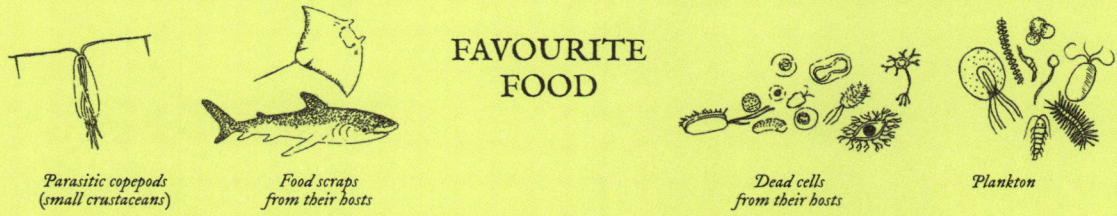

FAVOURITE FOOD

Parasitic copepods (small crustaceans)

Food scraps from their hosts

Dead cells from their hosts

Plankton

People used to believe that remoras fed on any food left over after their shark hosts had eaten.

However, it's now thought that they feed mainly on the parasites that live on the skin and gills of their host. In other words, the remoras get rid of the shark version of head lice. So the sharks get something from the relationship too!

WHEN DO THEY EAT?

As far as we know, they eat whenever they can.

WHO EATS THEM?

While attached to their host, they have no predators. Would you dare go near a shark to get a little bite of remora?

1. oesophagus
2. stomach
3. intestine
4. anus

The remora's dorsal fin works just like a suction cup.

HOLDING ON TIGHT

The remora's secret weapon is its dorsal fin, which looks like a suction cup. It is made up of sixteen to twenty soft rods called laminae. These move to create a vacuum and allow the remora to stick itself to a shark, or a turtle, or even the bottom of a boat.

A LIVING FISH HOOK

Some fishermen use remoras to catch fish: they tie them to a fishing line, cast it, wait for the remoras to attach themselves to another fish, and then reel the line in!

This Australian mammal has a bill like a duck, is venomous and lays eggs. It's one of a very small group of mammals called monotremes.

42

It looks so unusual that when European scientists examined the body of a platypus for the first time, they thought it was a hoax.

PLATYPUS

Ornithorhynchus anatinus

Animal:
VERTEBRATE

Digestive system:
MONOGASTRIC

Type of diet:
CARNIVORE

No acid, no problem

The digestive system of the platypus is unusual. Firstly, its stomach is much smaller and less developed than those of most mammals, and it does not produce gastric acids. As well as laying eggs like a bird, it also has a cloaca, which is a single hole it uses to pee, poo, and lay eggs.

FAVOURITE FOOD

Freshwater shrimp

Insect larvae

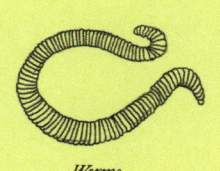
Worms

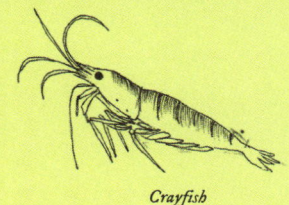
Crayfish

The platypus mainly eats small invertebrates that are high in protein, so it doesn't need a complicated stomach.

It is an excellent swimmer and hunts underwater by stirring up the bottom of rivers and lakes with its snout.

WHEN DO THEY EAT?

They spend around twelve hours a day foraging for food.

WHO EATS THEM?

Its natural predators include snakes, birds of prey, foxes, dogs, and traditionally human beings, who hunted it for its fur. It's now a protected species.

1 oesophagus
2 stomach
3 intestine
4 cloaca

Platypus poo comes out here. So do the eggs that the females lay!

Platypus don't have teeth. They do as babies, but then they lose them.

A NOSE FOR ACTION

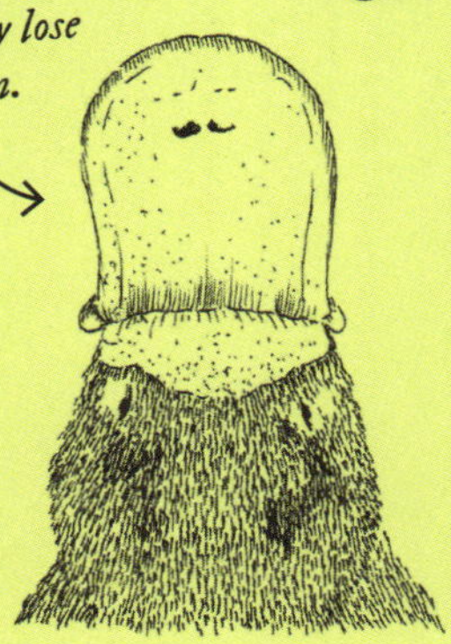

What looks like a duck's beak is actually the platypus' snout. As well as stirring up the river bed, it can detect scents underwater. It also contains electroreceptors, which the platypus can use to detect the movements of its prey.

A TAIL FOR HARD TIMES

A platypus tail is wide and flat like a beaver's tail. They use it to steer themselves through the water, and it is also used to store fat for times when food is scarce.

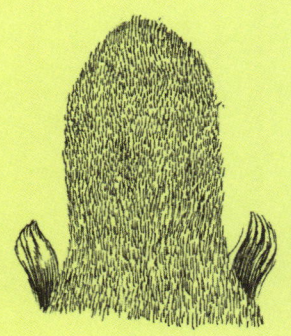

Wombat droppings are very dry because the wombat extracts all the water it can from its food. But the most fascinating thing is that they are cube shaped!

43

A wombat's digestion is extremely slow and can take up to two weeks. Making cubes takes a long time!

WOMBAT
Vombatidae

Animal: **VERTEBRATE** | Digestive system: **MONOGASTRIC** | Type of diet: **HERBIVORE**

Cubic poo!

If having a bowel movement were an art form, the wombat would be a great artist. It's the only animal that produces cubes of poo. Its droppings have this shape so that they don't roll around and are more likely to stay where the wombat puts them. This is important because wombats use poo to mark their territory and to attract potential mates.

FAVOURITE FOOD

Grass

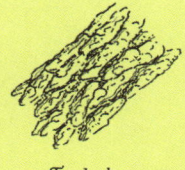
Tree bark

Shrubs

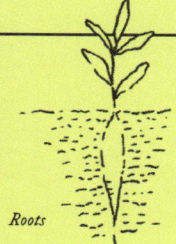
Roots

A wombat's diet is made up of plants with very few nutrients in them.

For this reason, its digestive system is really long — ten times longer than its body! This means it can extract every last bit of goodness from its food.

WHEN DO THEY EAT?

They are mostly nocturnal, although they also go out looking for food on cloudy and cold days.

WHO EATS THEM?

Their main predators are other Australian animals including the dingo (a type of wild dog) and the Tasmanian devil.

1. oesophagus
2. stomach
3. small intestine
4. anus

A wombat can produce up to 100 cubic droppings in a single day.

MAKING CUBES

How do wombats give their droppings a cubic shape? No, their anus isn't square, and they don't have a mould in their intestines! But the last segment of their large intestine is very stretchy, and this is where they shape their droppings.

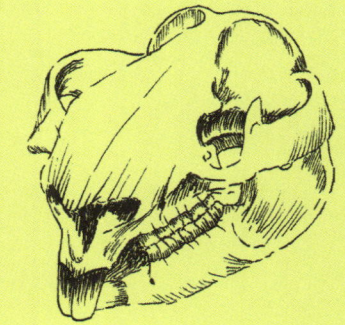

FRONT TEETH FOREVER!

Wombat are not rodents, but marsupials. However, their teeth are very similar to rodent teeth because of their diet. They have large rootless incisors (front teeth) that keep growing throughout their lives.

Dogs are the most popular pet in the world. It is estimated that there are nearly one billion dogs on the planet.

44

Although dog breeds can look very different, they all share the same genetic background and are part of the same subspecies.

DOG

Canis lupus familiaris

Animal:
VERTEBRATE | Digestive system:
MONOGASTRIC | Type of diet:
OMNIVORE

Shaped by humans

Between 14,000 and 29,000 years ago, human beings started to domesticate the wolf (*Canis lupus*). As a result of its new way of life, the domestic dog developed and became dependent on humans. Its diet changed from carnivorous and based on hunting to including other foods provided by humans, such as cereals. These led to changes in the dog's digestive system: its teeth got smaller, its jaws grew weaker, and its gut bacteria became more complex, in order to digest starches.

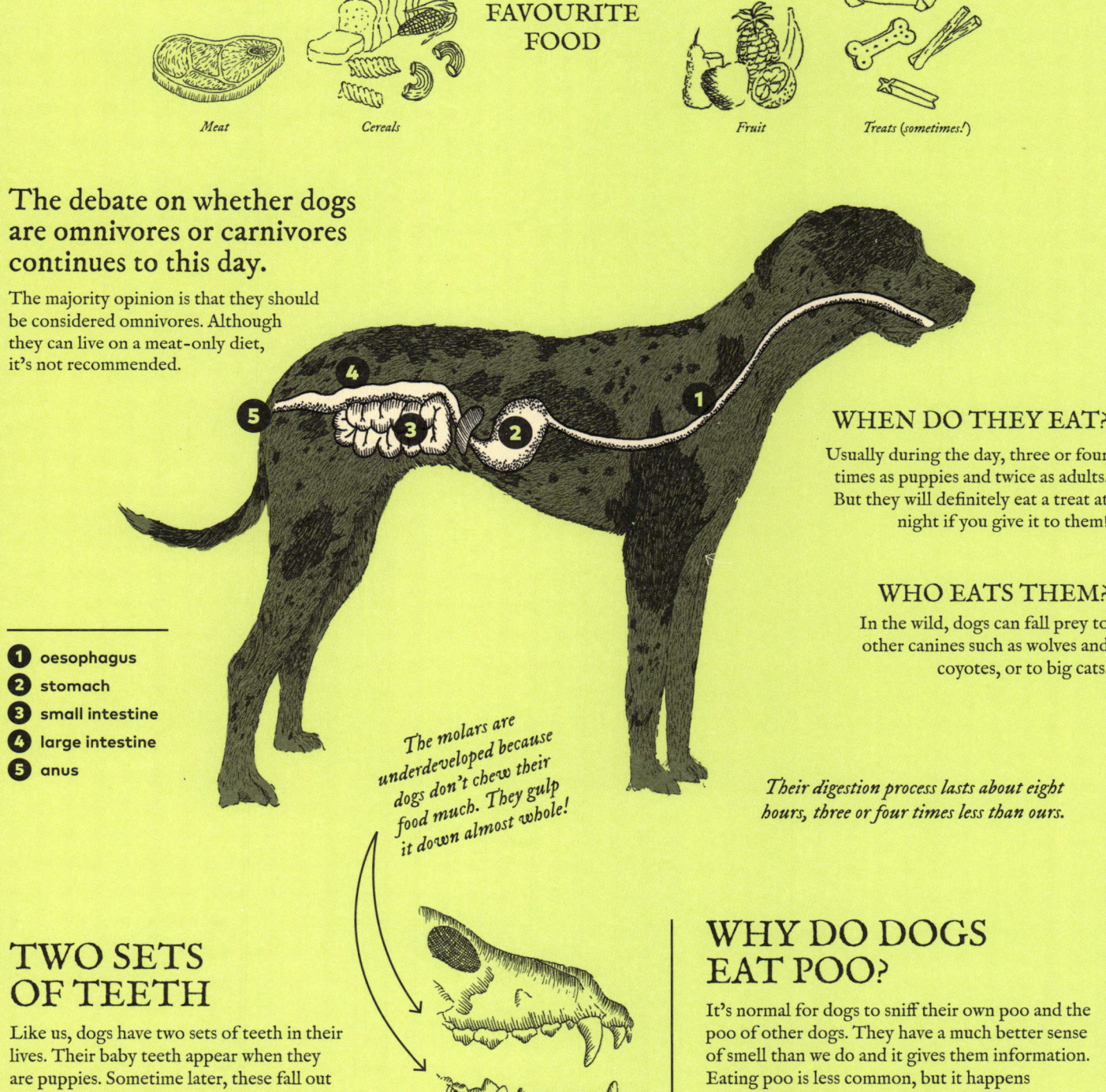

FAVOURITE FOOD

Meat *Cereals* *Fruit* *Treats (sometimes!)*

The debate on whether dogs are omnivores or carnivores continues to this day.

The majority opinion is that they should be considered omnivores. Although they can live on a meat-only diet, it's not recommended.

1. oesophagus
2. stomach
3. small intestine
4. large intestine
5. anus

The molars are underdeveloped because dogs don't chew their food much. They gulp it down almost whole!

WHEN DO THEY EAT?
Usually during the day, three or four times as puppies and twice as adults. But they will definitely eat a treat at night if you give it to them!

WHO EATS THEM?
In the wild, dogs can fall prey to other canines such as wolves and coyotes, or to big cats.

Their digestion process lasts about eight hours, three or four times less than ours.

TWO SETS OF TEETH

Like us, dogs have two sets of teeth in their lives. Their baby teeth appear when they are puppies. Sometime later, these fall out and their adult teeth grow. The first set has 28 teeth and the second set has 42.

WHY DO DOGS EAT POO?

It's normal for dogs to sniff their own poo and the poo of other dogs. They have a much better sense of smell than we do and it gives them information. Eating poo is less common, but it happens sometimes. It is usually a symptom of a diet problem or other changes in behaviour.

Like other pufferfish, it is a fairly poor swimmer. It's slow and can't move around very easily.

45

Fortunately, it has an excellent way of avoiding predators, by turning into a big spiky ball!

PORCUPINE PUFFERFISH
Diodon holocanthus

Animal: **VERTEBRATE** | Digestive system: **MONOGASTRIC** | Type of diet: **CARNIVORE**

All puffed up

The body of this fish, from its skeleton to its skin, has evolved to allow it to puff itself up. Its best defensive weapon is its digestive system, so much so that its stomach has lost some of its original functions. When pufferfish feel threatened, they swallow lots of water (or air, if a bird pulls them out of the sea) and then the magic happens: their stomach stretches and swells, expanding the peritoneal cavity, which is the space where their digestive organs are located. That makes them very hard to swallow!

 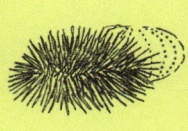 

FAVOURITE FOOD

Hermit crabs *Sea urchins* *Barnacles* *Sea snails* *Other shellfish*

Air, water and all kinds of shellfish move through the stomach of a porcupine pufferfish.

Since they are not very good swimmers, their diet is made up of animals that can't move quickly.

Handle with care!

WHEN DO THEY EAT?

They hunt and feed by night because of their good eyesight. By day, they rest in reefs or underwater caves and hollows.

WHO EATS THEM?

Only some large predators like orcas and sharks are able to tackle the thorny issue of how to eat these animals as adults.

1. digestive tube
2. stomach
3. small intestine
4. large intestine
5. anus

NUTCRACKING JAWS

The mouth of the porcupine pufferfish is similar to the beak of a parrot. The teeth on both the upper and lower jaw are joined together to form solid plates. These can easily break through the shells of the pufferfish's prey to get to the meat.

SPINES, NOT SCALES

The spines of this pufferfish have evolved from scales. Under normal conditions, they remain flat, lying close to the skin, but when the fish puffs up, they stand up straight and sharp, making the pufferfish look much too spiky to eat.

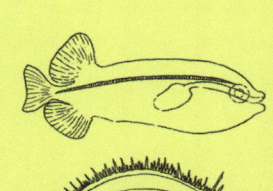
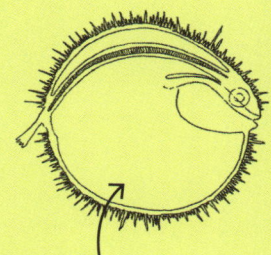

Porcupine pufferfish can triple in size when they swell up.

The common anaconda can reach 8 metres in length and weighs as much as a human adult. Along with the reticulated python, it is the largest snake in the world.

46

It is a constrictor snake, which means it kills its prey by wrapping itself around it and crushing it until it stops breathing.

COMMON ANACONDA
Eunectes murinus

Animal: **VERTEBRATE** | Digestive system: **MONOGASTRIC** | Type of diet: **CARNIVORE**

Down in one bite

Doctors recommend that you chew your food properly before swallowing it. But anacondas don't follow that advice! They 'dislocate' their jaw to swallow their prey whole without slicing or chewing it. They start with the prey's head and then gradually stretch their mouth over its body until it's completely swallowed. Their powerful digestive system takes care of the rest — it can even break down bones. The only thing it can't handle is keratin, the stuff that hair, claws and beaks are made of.

FAVOURITE FOOD: Capybaras, Tapirs, Peccaries, Deer, Monkeys, Birds, Fish, Bats

Anacondas can swallow prey weighing up to 40 or 50 kg.

In these cases, digestion can take weeks. During this time, the snake moves very little and devotes itself to digesting.

1. oesophagus
2. stomach
3. liver
4. intestine
5. anus

WHEN DO THEY EAT?

They seem to be more active as night falls. When hunting, they track their prey underwater or pounce on it from the branches of trees.

WHO EATS THEM?

Their main predators are jaguars and alligators, although these only dare to attack younger or smaller anacondas.

Like many snakes, the anaconda has a huge liver. The substances it produces help with the snake's long and complicated digestive process.

The quadrate bones in the anaconda's jaw are very long. They allow the snake to open its mouth extremely wide to swallow its prey.

180 degrees

TEETH ABOVE...

Anacondas have more than a hundred teeth, grouped together in four rows. Two of these rows are in the upper jaw. These teeth are used to grab on to their prey as they coil themselves around it.

...AND BELOW

The two rows of teeth on the lower jaw, on the other hand, are the ones they 'stretch' around the body of their prey before gulping it down whole.

In the wild, this adorable animal is found only in a few mountainous areas of central China.

47

The giant panda has become a symbol in the fight to protect the world's endangered animals.

GIANT PANDA

Ailuropoda melanoleuca

Animal: **VERTEBRATE** | Digestive system: **MONOGASTRIC** | Type of diet: **HERBIVORE**

A herbivore with a carnivore's insides!

Meat is an essential part of the diet in nearly all species of bear. The giant panda is an exception as it is an herbivore. However, its digestive system is not well adapted to this vegetarian diet and it has a hard time extracting nutrients from the cellulose of plants. To top it all off, its favourite plant, bamboo, is quite low in nutrients. So what does it do? It just keeps eating! Pandas spend ten to fourteen hours a day feeding themselves with kilos and kilos of bamboo to meet their needs.

FAVOURITE FOOD

Bambo — *Fruit* — *Small mammals* — *Fish* — *Insects*

Bamboo makes up 99% of a panda's diet. It will eat other foods, but only occasionally.

Bamboo isn't very nutritious but it's easy to find and grows all year round. Pandas are low maintenance in this respect!

WHEN DO THEY EAT?

They spend practically all day eating, during daylight hours. By night, they rest in the bamboo forests.

WHO EATS THEM?

They have no natural predators. Human beings are their only possible threat, although they are now a protected species.

1. oesophagus
2. stomach
3. intestine
4. anus

The inside of a panda's intestines is coated with a thick layer of mucus to protect it from bamboo splinters.

Give me six!

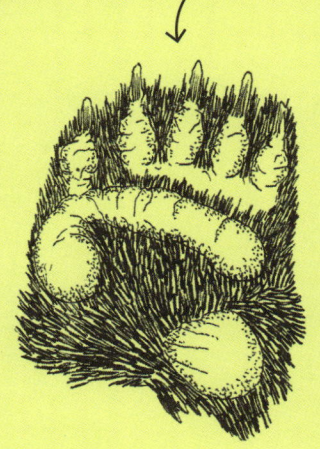

TEETH FOR CHEWING

The giant panda's teeth have adapted better to its favourite food than its stomach has. Its molars and premolars are wider and flatter than in other species of bear, which allows it to mash up the bamboo canes and reach the tastiest part of the plant, the pulp.

A SIXTH FINGER

Another part of the panda's body that has adapted to its diet is its paws. Pandas have a 'sixth finger', which is not really a real digit but a bone that forms part of their wrist and which has evolved into a longer shape. They use this thumb-like digit to hold bamboo.

It is thought that horses were first domesticated by human beings in around 3000 BCE.

48

An average horse eats about ten kilos of food a day and drinks more than forty litres of water.

HORSE

Equus ferus caballus

Animal:
VERTEBRATE

Digestive system:
MONOGASTRIC

Type of diet:
HERBIVORE

One-way traffic

The digestive system of horses is a one-way street. Not because they are not ruminants, but because the entry valve to their stomach is so strong that it cannot open in the opposite direction, which is what happens with most animals (like us). This means that horses can't burp or vomit, so they need to be careful about what they eat to avoid problems. They tend to eat little and often.

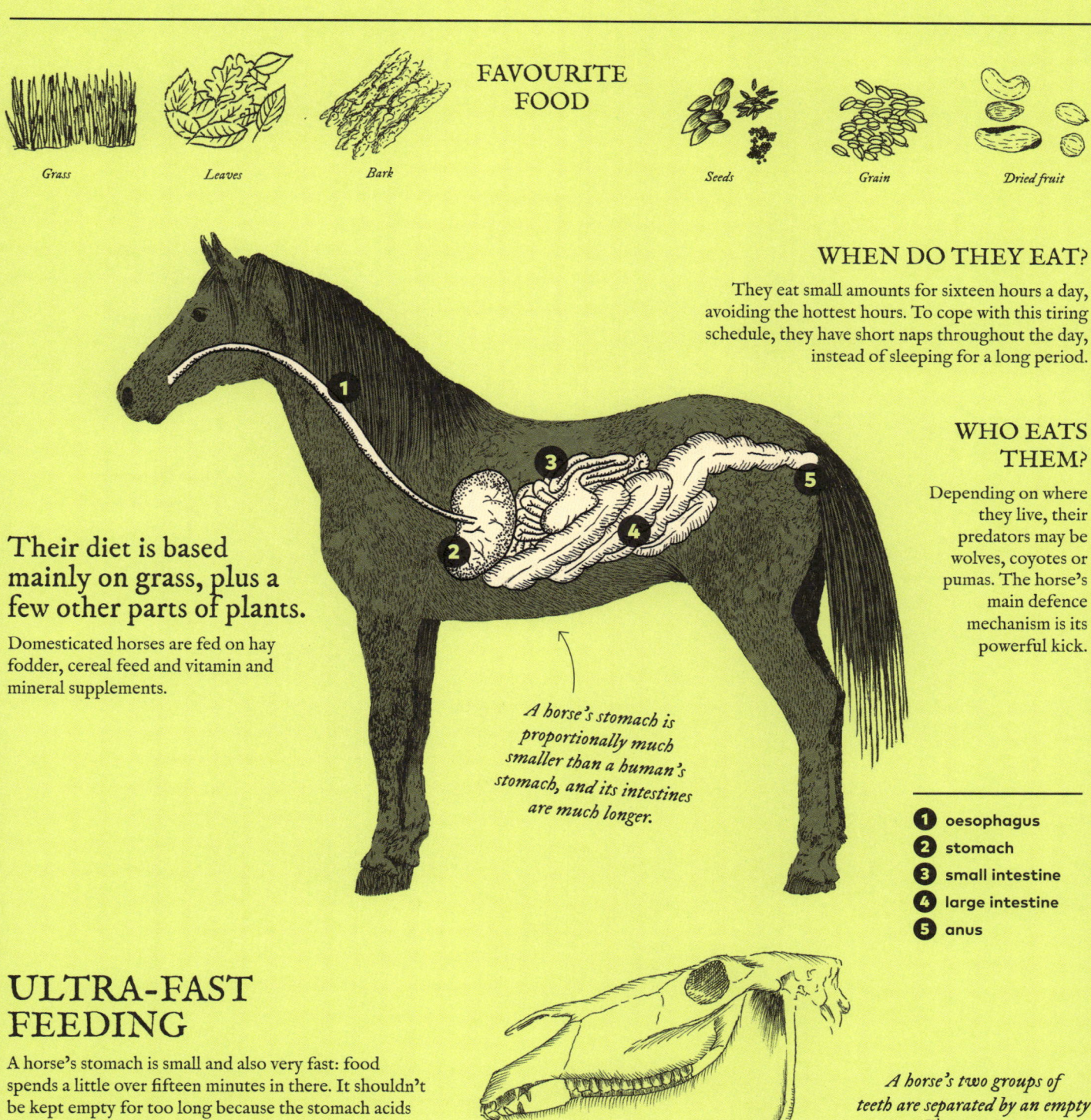

FAVOURITE FOOD

Grass · Leaves · Bark · Seeds · Grain · Dried fruit

WHEN DO THEY EAT?

They eat small amounts for sixteen hours a day, avoiding the hottest hours. To cope with this tiring schedule, they have short naps throughout the day, instead of sleeping for a long period.

WHO EATS THEM?

Depending on where they live, their predators may be wolves, coyotes or pumas. The horse's main defence mechanism is its powerful kick.

Their diet is based mainly on grass, plus a few other parts of plants.
Domesticated horses are fed on hay fodder, cereal feed and vitamin and mineral supplements.

A horse's stomach is proportionally much smaller than a human's stomach, and its intestines are much longer.

1. oesophagus
2. stomach
3. small intestine
4. large intestine
5. anus

ULTRA-FAST FEEDING

A horse's stomach is small and also very fast: food spends a little over fifteen minutes in there. It shouldn't be kept empty for too long because the stomach acids could cause problems such as ulcers. So they have to eat often for the good of their health!

A horse's two groups of teeth are separated by an empty space. This is where the bit of a harness is fitted.

The rabbit looks cute but it is considered one of the most damaging invasive species in the world.

49

Yes, rabbits do eat carrots, like you see in cartoons, but that's definitely not their favourite food!

EUROPEAN RABBIT
Oryctolagus cuniculus

Animal:
VERTEBRATE

Digestive system:
MONOGASTRIC

Type of diet:
HERBIVORE

Double digestion

The advantage that ruminant animals have over herbivores with a single stomach is that they can 'process' their food twice. The rabbit has found an unusual way of doing the same thing with its single stomach. When the food it has eaten is excreted as poo, the rabbit eats it again, sending it through its system for a second time and extracting the nutrients left over from last time. It may sound disgusting, but it works for them!

FAVOURITE FOOD

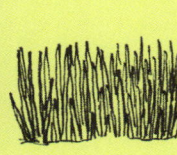  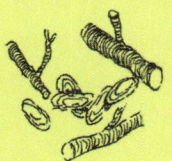   

Grass   *Stalks*   *Bark from shrubs*   *Roots*   *Seeds*   *Dried fruits*

Adult rabbits can eat up to half a kilo of food a day.

The plants they eat aren't very nutritious, so they need to eat a large amount.

1. oesophagus
2. stomach
3. small intestine
4. large intestine
5. anus

WHEN DO THEY EAT?

They eat at night, from sunset to sunrise. Then they spend nearly all day resting.

WHO EATS THEM?

They have a lot of predators including foxes, dogs, weasels, wild cats, eagles, owls and more. However, there are still plenty of them around because they breed... well, like rabbits!

In non-ruminant herbivores, the caecum (the first section of the large intestine) is large because this is where the toughest parts of plants are digested.

Rabbits look as if they have four incisors, but they actually have six. There are two more hidden behind the top two!

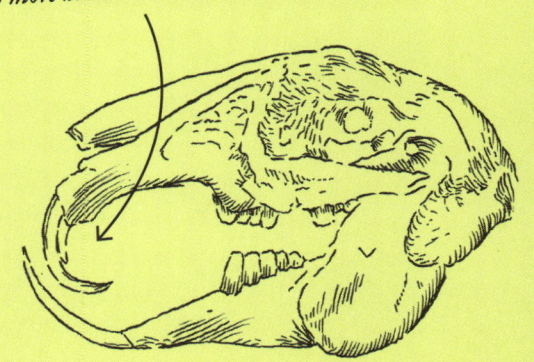

A TOOTHY PROBLEM

Rabbits need their teeth to eat but their teeth can also be a source of problems. Their teeth continue to grow throughout their lives, but a rabbit may not be able to feed itself properly if its large incisors grow too long. Fortunately, chewing wears down the teeth naturally.

TWO TYPES OF POO

After the initial digestion, rabbit poo is soft and full of bacteria and protein. These pellets are called 'caecotrophs', and the rabbits eat them again. The pellets they poo the second time are drier and darker.

Giant anteaters have a distinctive appearance. Their closest relatives are armadillos and sloths.

50

This is the largest anteater species. It can reach a length of two metres, although a lot of that is its tail!

GIANT ANTEATER

Myrmecophaga tridactyla

Animal: **VERTEBRATE** | Digestive system: **MONOGASTRIC** | Type of diet: **MYRMECOPHAGE**

Not all ants...

The name might make you think that giant anteaters only eat ants, but in fact they eat termites too. They eat both types of insect in the same way – first by finding a nest with their incredible sense of smell (as their sight is not very sharp), then opening it with their powerful claws and inserting their long, narrow snout. Then they use their ultra-long tongue to feast!

FAVOURITE FOOD

 *Ants* 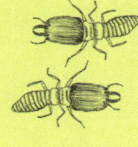 *Termites*  *Insect larvae*  *Fruit*

Anteaters feed on insects that are high in protein and easy to digest, so their digestive system is relatively simple.

They can eat tens of thousands of insects in only one day!

WHEN DO THEY EAT?

They are normally diurnal. However, when they live in heavily populated areas, they prefer to go out and eat at night.

WHO EATS THEM?

Their greatest threats are pumas and jaguars, but they can defend themselves pretty well. The biggest threat to their species is the destruction of the habitats they live in.

Along with ants and termites, they take in nearly 100 grams of soil a day, which which comes out in their poo.

1. oesophagus
2. stomach
3. small intestine
4. large intestine
5. anus

No one can stick their tongue out like an anteater!

A TONGUE LIKE LIGHTNING

An anteater's tongue is very long and thin. It can be up to 60 cm long and is ideal for pushing inside an ants' nest or termite mound. It has a spiny texture and is covered with sticky saliva to catch as many insects as possible with each flick of the tongue. These flicks are lightning fast – up to 150 times per minute!

CLAWS FOR A CAUSE

The formidable claws on their feet are especially long on their forelimbs. They don't just use them to open termite mounds and ants' nests. They are also great for self-defence. Anteaters stand up on their hind limbs when threatened and lash out with their forelimbs!

This is one of only three bat species that feed purely on blood. The others are the hairy-legged vampire bat and the white vampire bat.

51

Their bites are not actually dangerous but they can pass on dangerous diseases such as rabies to cattle and even to human beings.

COMMON VAMPIRE BAT
Desmodus rotundus

Animal:
VERTEBRATE

Digestive system:
MONOGASTRIC

Type of diet:
HAEMATOPHAGE

Blood feast!

Blood is not a very nutritious food. It contains a lot of protein, but very few carbohydrates and fats. This means that it provides vampire bats with very limited energy. They make up for this by feeding on blood often, and in large quantities. When they feed, they can eat up to one and a half times their own weight in blood. It's like a human drinking dozens of litres of water at once! And they need to do this almost daily. A bat that doesn't eat for two or three days could easily die.

FAVOURITE FOOD

Blood

These bats usually drink the blood of farm animals, such as horses, cows, pigs and sheep, because they make the easiest prey.

But they can also feed on wild animals and even humans if need be.

1. oesophagus
2. stomach
3. intestine
4. anus

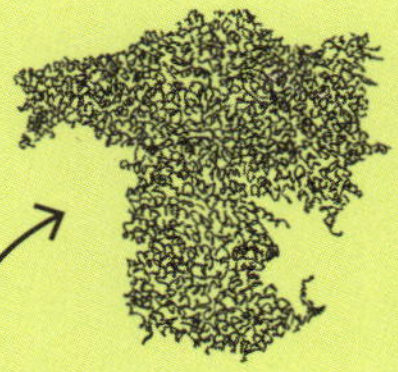

The poo of vampire bats is very liquid, as you might guess from their diet.

WHEN DO THEY EAT?

Not surprisingly, vampire bats are most active at night! By day, they use caves, empty packing crates or abandoned buildings as hiding places.

WHO EATS THEM?

Even vampire bats have vampire-hunters! In this case, their predators are usually birds of prey or snakes.

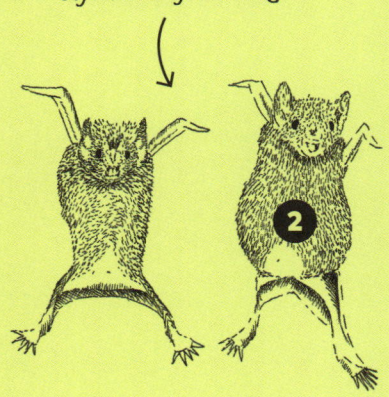

A vampire bat, before and after eating.

ONCE BITTEN

The upper incisors are the teeth these bats use for biting and they are razor-sharp. Once its teeth have pierced the prey, the bat licks up the blood coming out from the wound. The blood flows strongly because the vampire bat's saliva contains an enzyme that stops the wound from healing. For obvious reasons, scientists have named this substance Draculin.

A STRETCHY STOMACH

After feeding, the bellies of vampire bats are so full that they can't fly. For this reason, they absorb and expel the water in their food very quickly. Only two minutes after starting to eat, they start to pee. Even so, after they have finished eating, they are still too heavy to take off from the ground and fly. So they need to get airborne with a vertical leap, using their powerful legs to take off.

This species belongs to the order of tree shrews, one of the closest groups of mammals to the primates.

52

It lives in Borneo and south-east Asia, and like human beings, does its business in a toilet!

MOUNTAIN TREE-SHREW
Tupaia montana

Animal:
VERTEBRATE

Digestive system:
MONOGASTRIC

Type of diet:
OMNIVORE

A natural toilet

The mountain tree-shrew's best friend is the *Nepenthes lowii*. This is a type of pitcher plant, with a vase-like 'container' covered by a lid-like flap. Every morning, when the tree-shrew wakes up and needs to relieve itself, the plant opens its lid, which has tasty nectar on the inside. The tree-shrew sits on the opening to do its business, while eating the nectar. In exchange, the tree-shrew's poo provides the plant with nutrients it needs to survive.

FAVOURITE FOOD

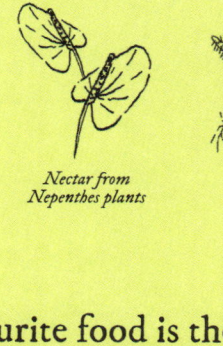
Nectar from Nepenthes plants

Insects and spiders

Fruit

Berries

Their favourite food is the nectar of the pitcher plant, which contains lots of sugar.

They get their protein from insects and other creatures that they hunt on the forest floor.

1. oesophagus
2. stomach
3. small intestine
4. large intestine
5. anus

WHEN DO THEY EAT?
Like almost all tree-shrews, they are diurnal.

WHO EATS THEM?
Their main predators are big cats and reptiles that live in the mountain forests of Borneo.

Their intestines are not able to digest fibre very well, so the plants in their diet are mainly fleshy fruits.

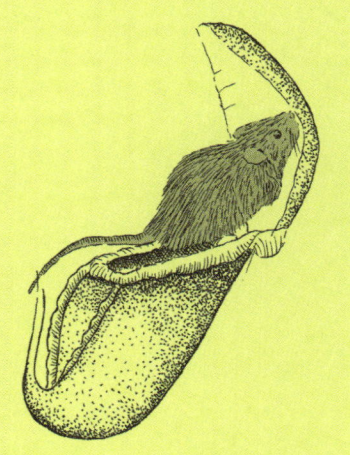

FROM CARNIVORE TO COPROPHAGE

You might think that *Nepenthes lowii* looks like a carnivorous plant and you wouldn't be far wrong. Carnivorous plants generally attract insects with their nectar, then trap them in some way. But *Nepenthes lowii* has evolved to 'eat' the tree-shrew's poo instead of bugs.

A IDEAL COUPLE

The two members of this partnership are made for one another: the aroma of the nectar attracts the tree-shrew and the colour inside the 'lid' catches its eye. These things encourage the tree-shrew to mark that location as its territory, using its poo.

Sloths sleep for up to eighteen hours a day and move veeeeery slooooowly.

53

The paws of these sloths have three toes, unlike some of their sloth cousins, which have only two.

THREE-TOED SLOTH
Bradypus tridactylus

Animal:
VERTEBRATE

Digestive system:
MONOGASTRIC

Type of diet:
HERBIVORE

Taking it slowly

The digestive process of sloths is the same as the rest of their lives, very slow. They can take up to thirty days to digest a leaf! It is the slowest digestion out of all mammals. This is because their diet is not very nutritious and they need time to extract all the nutrients from what they eat, using the bacteria that live in their digestive system. The metabolism of sloths is also extremely slow.

More than 99% of their diet is made up of leaves. Tender new leaves are their favourites.

They occasionally eat fruit or cacao beans.

FAVOURITE FOOD

Cecropia leaves   *Ceiba leaves*   *Cacao pods*

1. oesophagus
2. stomach
3. small intestine
4. large intestine
5. anus

WHEN DO THEY EAT?

It depends where they live: in colder regions, they eat during the hotter hours of the day, and in hotter regions, they eat at night.

WHO EATS THEM?

They are preyed on by harpy eagles, jaguars, wild dogs and humans.

Sloths spend a lot of time hanging upside down. For this reason, their oesophagus isn't a completely straight line. This way, they can swallow properly!

A sloth's weekly poo can be a dangerous time. It's much more vulnerable on the ground!

HIGH-RISE LIFE

These sloths live in the canopy of the lush Amazon rainforest. But once a week, they leave the safety of the trees and come down to the ground. What for? Well, to do their business! The reason why sloths pee and poo on the ground is not yet entirely clear.

MOTHS AND ALGAE

One idea is that it means the moths that live in the sloth's fur can lay their eggs in the sloth's poo. And what does the sloth gain from this? Well, it wants these moths to reproduce and live on it because they cause algae to grow in its fur. Algae gives the sloth a greenish colour, which is perfect for camouflage among the trees.

These cleaning fish live in colonies consisting of a male and several females. When the male dies, one of the females takes over, and transforms into a male!

They are small, measuring only 10 to 14 cm, but they live in close harmony with many large coral reef predators.

BLUESTREAK CLEANER WRASSE

Labroides dimidiatus

Animal: **VERTEBRATE** | Digestive system: **MONOGASTRIC** | Type of diet: **CARNIVORE**

A personal service

Cleaner fish don't keep the places they live in neat and tidy, but they do 'clean' parasites and dead skin off other species. These parasites are their main source of food. Because of this, even big fish like moray eels and groupers let these little guys sneak into their open mouths for some first-class dental hygiene, and come close to their eyes and gills for a skincare treatment.

FAVOURITE FOOD 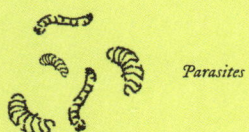 *Parasites*

Their diet is based on copepods and other invertebrates that live as parasites on their 'clients'.
They sometimes feed on crustaceans they catch themselves.

WHEN DO THEY EAT?
The bluestreak cleaning service is only open during the day. By night, they usually sleep on the seabed.

WHO EATS THEM?
Their 'clients' would be their natural predators. But the bluestreaks don't get eaten because the larger fish love the job they do!

1. oesophagus
2. stomach
3. small intestine
4. large intestine
5. anus

In one day's work, each cleaner wrasse can eat more than one thousand parasites.

CLEANING STATIONS

This is the name given to the places where these fish carry out their work. A small group, made up of one male and a few females or young fish, work at each station. When a client arrives, the cleaner fish perform a dance to 'ask permission' to start work. The client then adopts a relaxed position, which shows them that it is ready to be cleaned.

Sometimes there are even queues of fish waiting for their turn!

A FISHING SCAM?

Client fish recognize the bluestreak cleaner wrasse by its 'uniform', which is a dark horizontal stripe on each side of its body at eye level. However, there are often impostors in the neighbourhood! The false cleaner fish (*Aspidontus taeniatus*) is a similar-looking fish that infiltrates the stations and has learned the special bluestreak cleaner dance. But it doesn't clean. Instead, it eats the skin and flesh of the clients!

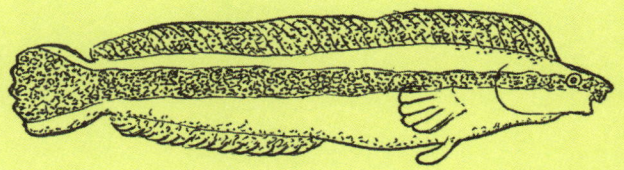

The blue whale is the largest animal that has ever existed on the planet, reaching 30 metres in length and weighing close to 200 tonnes.

55

This species eats two to four tonnes of food a day, which takes 16 hours to digest.

BLUE WHALE

Balaenoptera musculus

Animal:
VERTEBRATE | Digestive system: **MONOGASTRIC** | Type of diet: **PLANKTIVORE**

Small foods in big quantities

The open mouth of a blue whale can hold up to 90 tonnes of water and food. But it doesn't swallow it all! It filters out the water and eats only what it needs — krill, which is its favourite food. If you're wondering, this whale is not able to swallow a human being. Its oesophagus is too narrow, so stories about people ending up inside in a whale's stomach are impossible!

FAVOURITE FOOD

Krill

Their diet consists of krill, a small crustacean that's only a few centimetres long. It belongs to the family of zooplankton.

As they swallow krill, they also gulp down other passing animals like copepods and small fish.

WHEN DO THEY EAT?

By day, they eat at depths of 100 metres or more, and by night, they doze but will also nibble on a bit of krill at shallower depths.

WHO EATS THEM?

The only animals that dare come close to these big guys are orcas and humans. The whaling industry has has put them at risk of extinction.

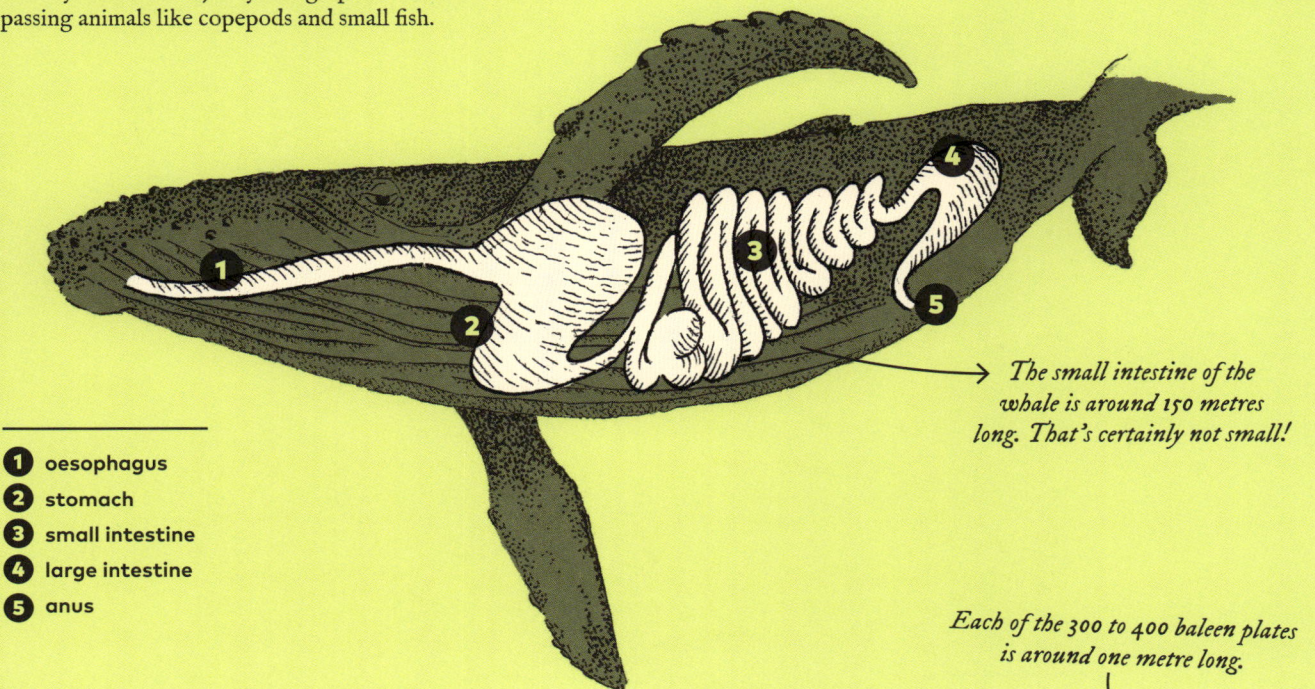

① oesophagus
② stomach
③ small intestine
④ large intestine
⑤ anus

The small intestine of the whale is around 150 metres long. That's certainly not small!

ONE STOMACH OR THREE?

Whales evolved from prehistoric land animals with hooves, the same group that many ruminants and pseudo-ruminants belong to. That is why, although they are not considered ruminants, their stomach is divided into several chambers. The first chamber is more or less the same as a ruminant's rumen, the second is the 'true' stomach where acids and digestive enzymes are produced, and the third helps to pump food through the system.

A FILTER FOR FOOD

Blues whales are part of a group known as baleen whales. Instead of teeth, they have a series of comb-like plates called baleen, covered in short hairs. They use the baleen to filter water from their mouths, so the water goes out but the krill stays in!

Each of the 300 to 400 baleen plates is around one metre long.

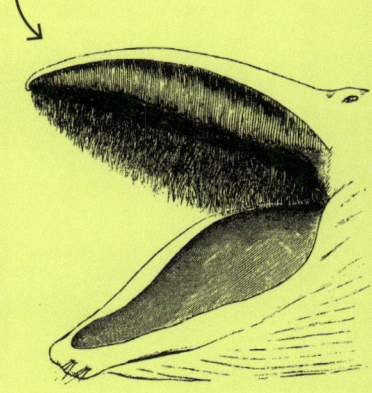

Hagfish are jawless fish. Their mouth is on the underside of their head and they have a rough, rasping tongue.

56

They have an extremely slow metabolism, so they can live for months without feeding.

HAGFISH

Myxinidae

Animal:
VERTEBRATE

Digestive system:
MONOGASTRIC

Type of diet:
CARNIVORE

An unusual way to eat

The digestive system of the hagfish is unique. The skin and intestine of these fish are slightly permeable — this means that they have the ability to soak up nutrients (especially amino acids, the molecules that proteins are made of) through their skin. It could be said that they eat through their skin. Just imagine if you could eat a cake by sticking your finger in it!

Fish carrion

Sea mammal carrion

FAVOURITE FOOD

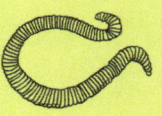
Small invertebrates

Protozoa

Hagfish feed on dead or dying animals.

For example, they love to feast on whale carcasses that sink to the bottom of the ocean.

WHEN DO THEY EAT?
They are normally nocturnal, especially if they live close to the surface. By day, they bury themselves in the seabed.

WHO EATS THEM?
Their slimy mucus protects them from most predators. But sometimes that's not enough to protect them from sablefish, mammals such as seals, or human beings.

① pharynx
② oesophagus
③ intestine
④ anus

In a few fractions of a second, a hagfish can produce about twenty litres of slime.

They have a very primitive digestive system. Most of it is a long intestine that works like a stomach, liver and pancreas, all at the same time.

SLIME TIME

Hagfish use slime to protect themselves. When they are attacked by another animal, their skin secretes very thin fibres of protein. At they come into contact with water, they turn into a mucus that clogs the mouth and gills of their attacker. The predator has to release its prey if it doesn't want to drown.

KNOT TODAY!

Hagfish love to tie their long and slender bodies into knots. They use this skill to latch onto a piece of carrion with their mouths and tear the flesh, making an opening so they can reach the meat and eat it from the inside. They also use this knot technique to clean off their slime or escape from predators.

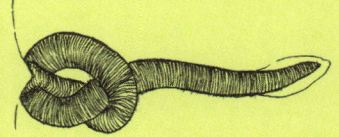

Pigs are considered one of the most intelligent species in the animal kingdom, even more so than dogs!

57

Despite their reputation, pigs are very clean animals. They only roll around in mud to cool down because their bodies can't sweat.

DOMESTIC PIG

Sus scrofa domestica

Animal:
VERTEBRATE

Digestive system:
MONOGASTRIC

Type of diet:
OMNIVORE

140

Eating like a pig

Sometimes humans eat like pigs, but we always digest like them. Our digestive systems have a lot in common: a single-chambered stomach, organs of a proportionally similar size, and a very similar way of sending food through the digestive tract. At the end of the day, we're both adaptable omnivores, although pigs eat more plants than we usually do.

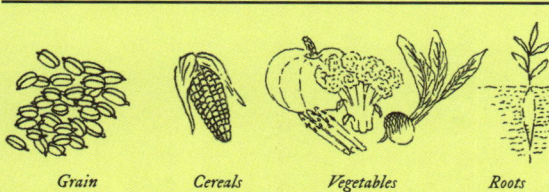

FAVOURITE FOOD

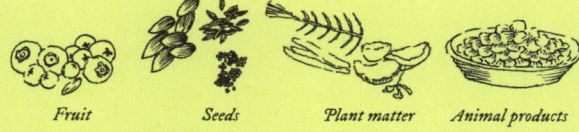

Grain   *Cereals*   *Vegetables*   *Roots*   *Fruit*   *Seeds*   *Plant matter*   *Animal products*

Like us, pigs are often happy with whatever food is available.

Their diet is based on vegetables, fruit, grains and cereals, but they will also eat insects, worms, eggs, small rodents or animal sub-products provided by farmers.

WHEN DO THEY EAT?

They are diurnal. Pigs that graze freely in the countryside can eat whenever they want. Pigs reared intensively in farms are usually fed twice a day.

WHO EATS THEM?

We human beings!

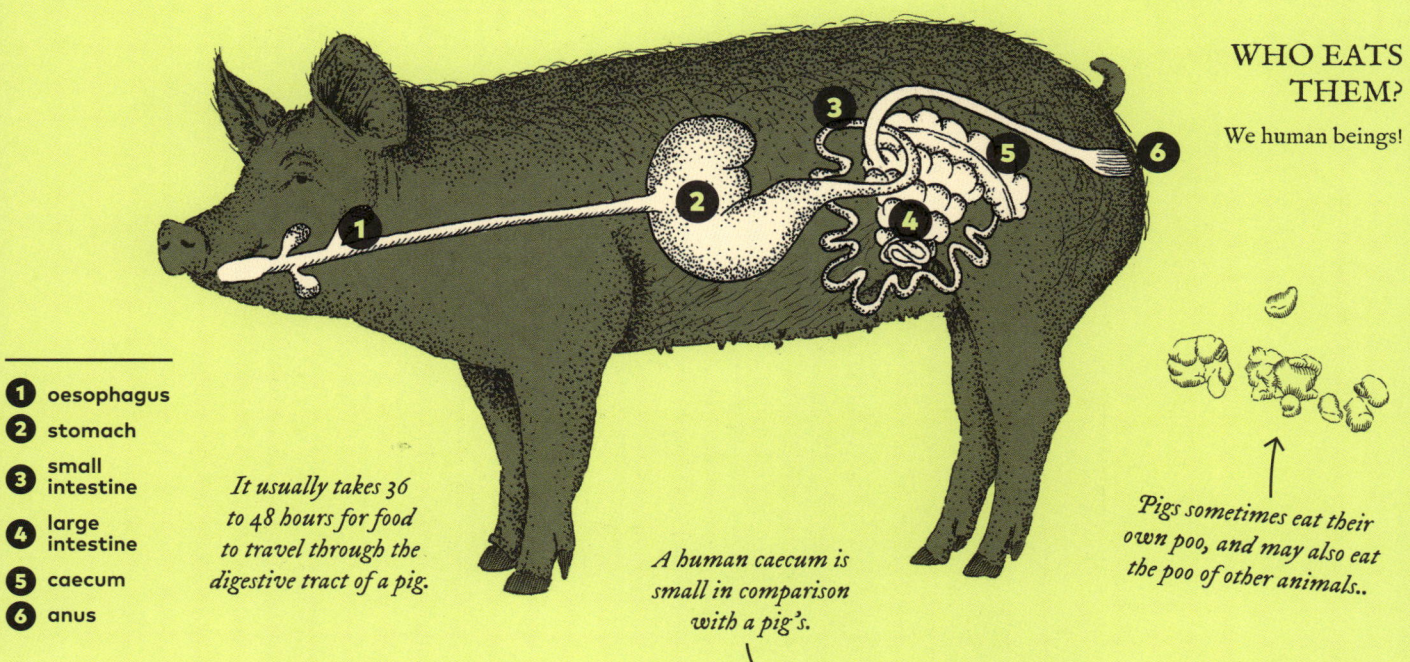

1. oesophagus
2. stomach
3. small intestine
4. large intestine
5. caecum
6. anus

It usually takes 36 to 48 hours for food to travel through the digestive tract of a pig.

A human caecum is small in comparison with a pig's.

Pigs sometimes eat their own poo, and may also eat the poo of other animals..

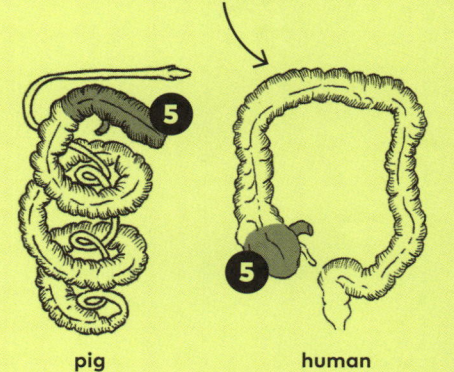

pig   human

THE CAECUM

What makes a pig's digestive system different from our own is the caecum, which is far bigger than in humans. This pouch at the entrance to the large intestine plays an important role in digesting the high-fibre foods that pigs eat. Humans eat vegetables with less fibre that are almost always cooked, so the caecum is smaller and doesn't have to work as hard.

DO PIGS EAT POO?

Pigs do not regularly eat their own poo, unlike it is for rabbits for example. However, it is true that they may eat it occasionally for nutrients when their diet is deficient.

Chimpanzees, along with bonobos, are the closest relatives of humans. They share more than 98% of their DNA with us!

58

Chimps are very social animals that live in communities with dozens of members.

CHIMPANZEE
Pan troglodytes

Animal:
VERTEBRATE

Digestive system:
MONOGASTRIC

Type of diet:
OMNIVORE

Planet of the apes

You probably use cutlery to eat, but it's very rare for animals to do anything like that. Chimpanzees, however, are very good at using tools — although not as good as us, of course. They use branches to get termites out of nests, stones to open hard-shelled fruit, and can even sharpen sticks with their teeth and use them to hunt, like a spear. These skills make feeding much easier!

FAVOURITE FOOD

Ripe fruit — *Young leaves* — *Seeds* — *Insects* — *Eggs* — *Smaller primates*

Although we tend to consider them omnivores, chimps prefer plant-based food. Fruit makes up more than 60% of their diet.

The animals they eat most often are insects, but they will sometimes feed on bigger animals, such as the red colobus monkey. Interestingly, female chimps eat less meat!

WHEN DO THEY EAT?
They are most active during the day, rather like humans.

WHO EATS THEM?
Leopards, pythons and crocodiles try to hunt them, although generally they prefer not to, because groups of chimpanzees are good at defending themselves.

1. oesophagus
2. stomach
3. small intestine
4. large intestine
5. anus

Chimps sometimes eat the bitterleaf plant (Vernonia amygdalina) when they have a parasite infection.

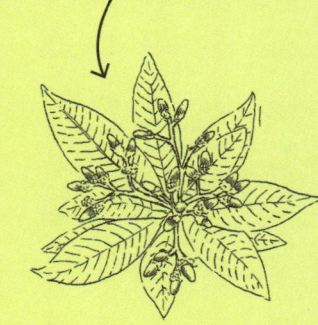

chimpanzee — human

THUMBS UP!
Like many primates, chimpanzees have opposable thumbs on both their upper and lower limbs. This allows them to use tools to feed themselves (or to hunt). The major difference is that chimpanzee thumbs are shorter than ours.

CHIMP MEDICINE
Chimps have been known to eat least thirteen types of medicinal herbs when they are unwell. In many cases, these are the same herbs used by people in local villages to treat diseases such as parasitic infections, stomach aches or headaches.

Lions are the biggest of the big cats after the tiger. Unlike most cats, they are social and usually live in small groups called prides.

59

LION

Panthera leo

They are not very active animals, and spend about twenty hours a day resting.

Animal:
VERTEBRATE

Digestive system:
MONOGASTRIC

Type of diet:
CARNIVORE

Hunting is a job for girls

A pride of lions consists of several females, some cubs and younger lions, and a pair of males. The males are the ones with manes, but the lionesses are the ones who hunt as a group and are in charge of bringing the food home. And although they are smaller, they are also more agile and faster (at up to 60 km/h) than the males. The male lions' excuse is that their manes cause them to overheat when they run around. What a lazy lot!

FAVOURITE FOOD

Wildebeest  *Zebra*  *Impala*  *Gazelles*

Lions sometimes attack very large prey, such as buffaloes and giraffes, but prefer to avoid them because their chances of success are poor.

At other times, they may feed on carrion, competing with hyenas and vultures.

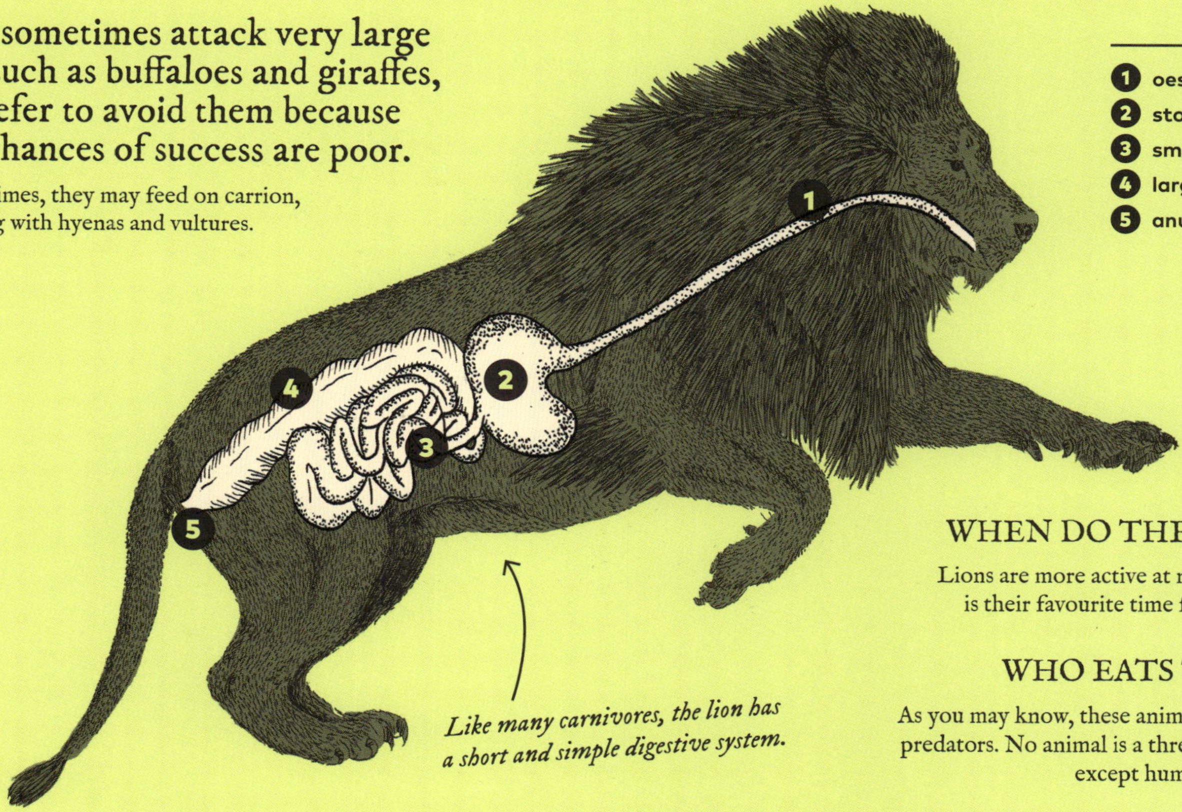

1. oesophagus
2. stomach
3. small intestine
4. large intestine
5. anus

Like many carnivores, the lion has a short and simple digestive system.

WHEN DO THEY EAT?

Lions are more active at night. Dawn is their favourite time for hunting.

WHO EATS THEM?

As you may know, these animals are apex predators. No animal is a threat to them, except human hunters.

VERY FELINE CANINES

Lions have huge canine teeth, which are about 8 cm long. In reality, they are not very useful for eating but they give impressive results when hunting. Lions use them to hold onto their prey until it suffocates or is strangled.

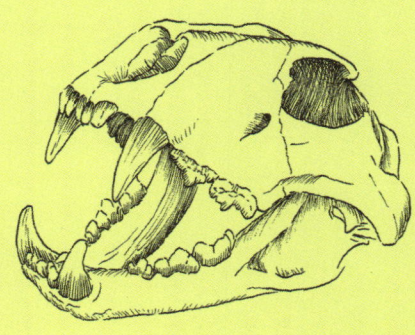

SHEARING TEETH

Another distinctive feature of a lion's mouth is the shearing teeth, also called carnassial teeth. Humans and herbivores have flat molars, which are used for chewing and grinding, but carnivores use their sharp shearing teeth for tearing and cutting meat.

Also known as the sewer rat, this rodent spread from China to all the world's continents, except Antarctica.

60

Rats are extremely adaptable, and can live in any kind of habitat except deserts or glaciers.

BROWN RAT

Rattus norvegicus

Animal: **VERTEBRATE** | Digestive system: **MONOGASTRIC** | Type of diet: **OMNIVORE**

The ultimate omnivore

A study of the diet of brown rats revealed the remains of 4,000 different food items in their stomachs. In fact, one of the keys to the success of these rodents is their ability to digest a huge variety of foods. Their ideal environment is big cities, where human food waste provides an almost endless source of food for them, so they're right at home.

FAVOURITE FOOD

 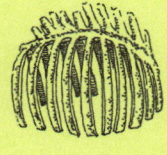   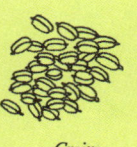 

Leftover human food — *Carrion* — *Small animals* — *Eggs* — *Grains and cereals* — *Fruit and vegetables*

In towns and cities, rats feed on garbage and food stored by humans.

In rural areas, their diet is based more on nuts and roots and small animals they either hunt or find dead.

WHEN DO THEY EAT?
They are nocturnal animals that go out and about at night to look for food, in places where they have learned it is easy to find.

WHO EATS THEM?
Carnivorous mammals, birds of prey and reptiles. We humans also kill them, trying to get rid of them because we consider them vermin.

1. oesophagus
2. stomach
3. small intestine
4. large intestine
5. anus

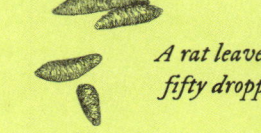

A rat leaves around forty to fifty droppings in one day.

EXTREME EATING

The diet of rats is so varied, it even includes things you may find rather disgusting. Firstly, they are coprophages, which means they eat poo. Although they do not do this as often as rabbits do, they may eat some of their own droppings to extract more nutrients from them. They may also take up cannibalism. If food is scarce, they will even eat their own babies!

ONE STOMACH OR TWO?

The stomachs of rats have adapted to 'binge eating' because they often need to eat in a hurry. Although the rat is a monogastric animal, the first part of the chamber (called the forestomach or non-glandular stomach) can store food for a few hours and pass small amounts (depending on the rat's need for energy) on to the next part of the stomach where the gastric juices are produced.

61.
Java mouse-deer

62.
Cow

63.
Giraffe

64.
Domestic sheep

65.
Red deer

66.
Iberian ibex

67.
Hippopotamus

68.
Bactrian camel

69.
Llama

70.
Vicuña

VERTEBRATES

Ruminants

VERTEBRATES

Pseudo-ruminants

Don't be deceived by its name: this is not a mouse! It is the world's smallest hoofed mammal.

61

As its name suggests, this species lives in the tropical forests of the Indonesian island of Java.

JAVA MOUSE-DEER
Tragulus javanicus

Animal: **VERTEBRATE** | Digestive system: **RUMINANT** | Type of diet: **HERBIVORE**

150

A primitive ruminant

Mouse-deer are considered the oldest members of the ruminant family. In fact, their primitive stomach only has three chambers: the rumen, the reticulum and the abomasum. The fourth chamber that ruminants have, the omasum, is not fully formed in them. Not only are they a living fossil, they're also very small — less than half a metre long.

FAVOURITE FOOD

Leaves

Shrubs

Shoots

Fungus

Fruit

The mouse-deer feeds on plants and fruits it finds on the ground in the lush forests where it lives.

In captivity, it also feeds on insects.

WHEN DO THEY EAT?
These little animals are usually more active at night.

WHO EATS THEM?
Because they are small, they are ideal prey for carnivorous mammals, birds and reptiles in the tropical rainforests of southeast Asia.

1. oesophagus
2. stomach (with three chambers)
3. intestine
4. anus

It's unusual for a deer to have fangs!

Its largest stomach chamber is the rumen, which makes up almost 90% of the stomach's capacity. It is proportionally bigger than the rumen of other ruminants.

NO TO HORNS, YES TO CANINES

They may be called deer but mouse-deer don't have antlers or horns. Instead, the males have a pair of tusk-like upper canine teeth. They don't use these for eating, but for defending themselves against enemies. Yes, they're tiny but they are fierce! Their lower jaw, however, has no canines.

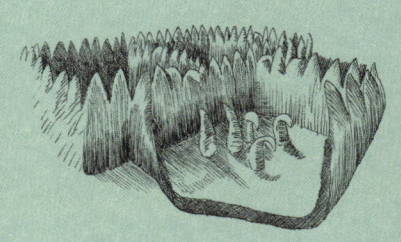

THE RETICULUM

Like other ruminants, the inside of the reticulum is formed by rows of honeycomb-like cells, with an inner structure similar to the one in the picture. This is where food particles are sorted: the largest go back to the rumen, while the smaller ones continue their journey through the digestive tract.

Cows can eat about 70 kg of fresh grass in eight hours. They do this every day.

They chew 40,000 times a day: 10,000 times when eating and 30,000 times when chewing the cud.

COW
Bos taurus

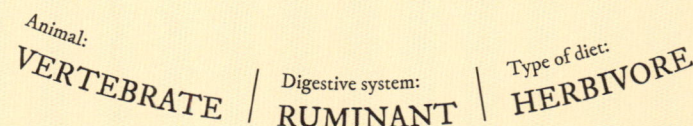

Animal:
VERTEBRATE | Digestive system:
RUMINANT | Type of diet:
HERBIVORE

Gas generator

Cows have many tiny friends to help them break down the grasses they feed on: millions of microorganisms (especially bacteria) live in their rumen and are responsible for this mammoth task. However, this process creates large quantities of gas called methane, a major cause of the greenhouse effect. So cows cause a lot of pollution!

FAVOURITE FOOD

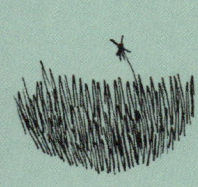  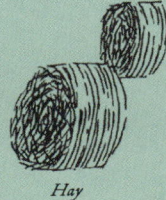    

Fresh grass   *Clover*   *Hay*   *Barley*   *Dandelions*   *Alfalfa*   *Oats*

Cows love to graze on fresh grass.

However, it is common for farmers to feed them dry fodder as well. Both are difficult and slow to digest, so the process can take up to 100 hours.

WHEN DO THEY EAT?

They like to graze in the cooler hours of the day. When the sun goes down, they prefer to graze in shady places, where the grass is colder.

1. oesophagus
2. stomach (with four chambers)
3. small intestine
4. large intestine
5. anus

WHO EATS THEM?

As cows are reared for food, their main predator is humans.

Cow manure is soft and paste-like. It can be used for compost, fuel or even for building houses.

The inner wall of a cow's rumen looks like a woollen rug!

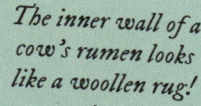

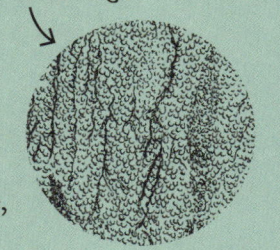

RUMINANT RUG

Some of the nutrients broken down in the rumen are absorbed through its walls. The walls are covered in thick strands called papillae, which help with this task. Other nutrients are absorbed at a later stage, as the food moves through the intestine.

SOMETHING TO CHEW ON

Like most ruminants, cows only have incisors in their lower jaw. The upper jaw has a hard pad instead of teeth. When grazing, they use their tongue to roll up a clump of grass, which they then cut with their lower incisors.

These ruminants can eat 30 to 60 kg of leaves a day. You might think this sounds like a lot, but it's not much for their size!

A giraffe's intestines are 80 metres long. Its neck doesn't seem quite so long in comparison!

63

GIRAFFE
Giraffa camelopardalis

Animal: **VERTEBRATE** | Digestive system: **RUMINANT** | Type of diet: **HERBIVORE**

High-rise feeding

Unlike most ruminants, giraffes feed on very nutritious plants, so in comparison with their size, they need to eat much less than other herbivores. They can do this because their long neck allows them to reach leaves that other animals can't get to, in the canopy of very tall trees. Their oesophagus has very strong muscles to help them regurgitate their food when ruminating.

FAVOURITE FOOD

Combretum bushes — *Mimosa leaves* — *Apricot tree leaves* — *Acacia leaves*

They love to eat the leaves from acacia trees, which can grow up to 12 metres in height.

Occasionally, they feed on grass, nuts and seeds. And if they're stressed, they chew bark!

1. oesophagus
2. stomach (with four chambers)
3. small intestine
4. large intestine
5. anus

WHEN DO THEY EAT?
They usually feed in the morning and evening. The rest of the time, they ruminate, standing upright by day and lying down at night.

WHO EATS THEM?
Their predators include lions, hyenas, leopards, wild dogs and crocodiles, but these will usually only attack giraffes that are sick, very old or very young.

Giraffe droppings look like little round balls.

TONGUE OUT!

The tongue of an adult giraffe is around 50 cm long and is prehensile, which means it can grab things. It allows giraffes to reach the high branches of acacia trees and grab the leaves. It is also covered in thick tissue that protects it against tree thorns.

It has a very striking colour, an almost blackish purple. This is because it contains a lot of melanin, a substance that helps protect it against sunburn. Giraffes spend hours eating with their tongue out, so it needs sun protection!

The sheep was one of the first animals to be domesticated, because it provided meat, milk and wool.

Sheep spend at least eight hours a day ruminating their partially digested food.

DOMESTIC SHEEP
Ovis orientalis aries

Animal: **VERTEBRATE** | Digestive system: **RUMINANT** | Type of diet: **HERBIVORE**

A rumen on its own

Like other ruminants, the first of the four chambers of a sheep's stomach is called the rumen and is the largest. In sheep, it can hold up to eight litres, which means they can take in a large amount of grass in a very short time. The rumen contains more than 350 species of bacteria, fungi and protozoa that break down the cellulose from grass in a very efficient way.

FAVOURITE FOOD

Pulses

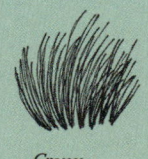
Grasses

Weeds

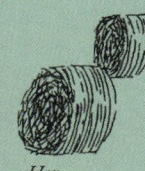
Hay

Sheep have a well-developed sense of taste, which means they prefer bitter-sweet plants to bitter ones.

In the winter, farmers feed them on hay (dried grass).

WHEN DO THEY EAT?

They are diurnal animals and more so morning ones. In the early hours of the day, they eat more, then they spend most of the rest of the day ruminating and digesting.

WHO EATS THEM?

Their predators vary depending on where they live: sometimes wolves, jackals and domestic dogs, but also cats, bears, birds of prey, and wild pigs.

1. oesophagus
2. stomach (with four chambers)
3. small intestine
4. large intestine
5. anus

MAGIC MANURE

Like other ruminants, the digestive process of sheep is particularly slow, taking up to 90 hours. The results are worthwhile because their manure is considered the best fertilizer. It is estimated that it is three times more effective than cow manure!

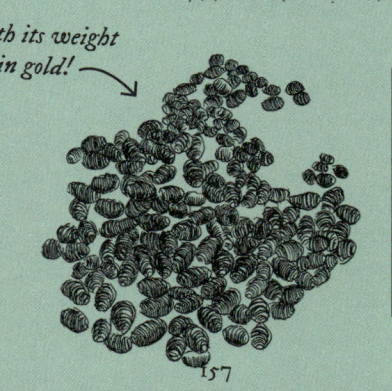

Worth its weight in gold!

PASS THE SALT

To keep their digestive system healthy, an adult sheep needs about nine grams of salt a day. But the pasture they feed on doesn't always contain enough minerals. This is why they sometimes lick naturally salty rocks. Farmers usually provide them with large blocks of salt to lick whenever they want.

After the moose and the elk, the red deer is the largest deer on the planet. It can reach two metres in length and about 200 kg in weight.

In Oceania and South America, the red deer was introduced as a game species for hunting, and has had a negative impact on many local ecosystems.

RED DEER

Cervus elaphus

Animal: **VERTEBRATE** | Digestive system: **RUMINANT** | Type of diet: **HERBIVORE**

Neither browsing nor grazing

Browsing animals are creatures that eat the leaves and tips of tree branches, and generally have a long large intestine. Grazing animals feed on pasture and grasses and have a comparatively long small intestine. Red deer are a mixture of the two: depending on the time of the year, they eat flowering plants and grasses or young tree branches. So the balance between the length of their small and large intestines lies somewhere in between the two groups.

FAVOURITE FOOD

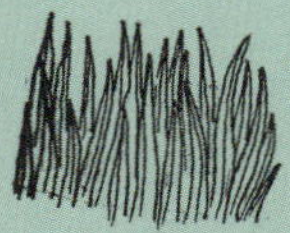

Grass    *Fungus*    *Flowers*    *Leaves*

Especially in the winter, deer eat a lot of young leaves. They also eat twigs and bark, which are hard to digest.

Fortunately, their four-chambered ruminant stomach allows them to get the most out of their food.

WHEN DO THEY EAT?

They are crepuscular animals, which means they are most active when the sun rises and sets. In the middle of the day and night, they rest.

WHO EATS THEM?

Depending on where in the world they live, adult red deer are preyed on by lynxes, wolves, grizzly bears, tigers and leopards. Smaller carnivores will often hunt younger deer.

1. oesophagus
2. stomach (with four chambers)
3. small intestine
4. large intestine
5. anus

Their antlers grow every summer and fall off in the winter.

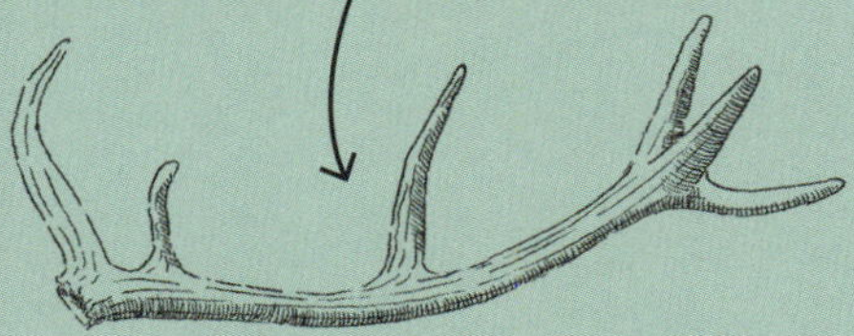

Their droppings are round, like small cannonballs.

FIGHTING SEASON

During the breeding season starting in the autumn, the males use up their energy fighting each other with their large antlers (they grow new ones every year) and mating with the females of the group if they win. If they don't build up enough fat reserves for winter, many die of hunger and exhaustion!

FOOD FILTER

The second stomach chamber of the deer (and ruminants generally) is responsible for stopping food that can't be digested from continuing down the digestive tract. Corn (which they don't digest well) and small stones, for example, have been found in the reticulum of deer.

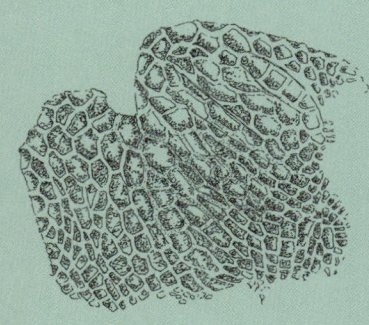

This species of goat is only found in the mountainous regions of Spain and Portugal.

The males are visibly larger than the females and also have much longer horns.

IBERIAN IBEX
Capra pyrenaica

Animal: **VERTEBRATE** | Digestive system: **RUMINANT** | Type of diet: **HERBIVORE**

A chamber for everything

The omasum is the third stomach chamber of the Iberian ibex (and ruminants generally). It is the last section of what is called the forestomach, which comes before the section known as the 'true stomach' or abomasum. The omasum works like a pump. It contracts and relaxes in order to push the food into the next chamber, while squeezing the liquid out of it. The omasum is also where nutrients such as fatty acids begin to be absorbed, along with water.

FAVOURITE FOOD

Boxtree,  Broom,  Thyme, 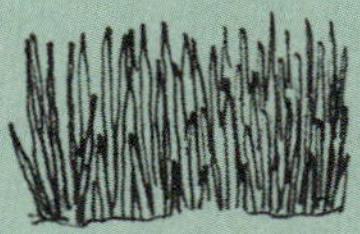 Grass,  Acorns,  Seeds, 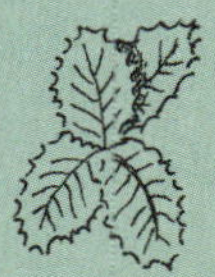 Evergreen oak,  Rosemary,  Weeds

The diet of an ibex changes with the seasons. In the cold months, it mostly eats branches, shoots and leaves from trees and shrubs.

In good weather, it prefers to feed on weeds and grass.

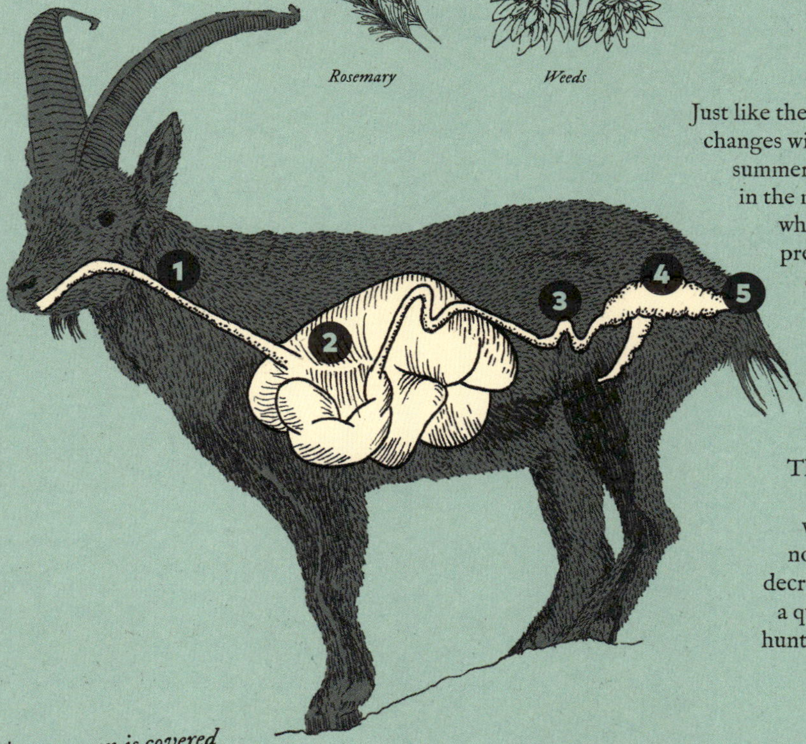

1. oesophagus
2. stomach (with four chambers)
3. small intestine
4. large intestine
5. anus

The omasum is covered with page-like flaps called laminae.

WHEN DO THEY EAT?

Just like their diet, their daily life changes with the seasons. In the summer, they are most active in the morning and evening, while in the winter, they prefer the middle of the day when it's warmer.

WHO EATS THEM?

Their natural predators used to be bears and wolves, but these have now died out or greatly decreased. The ibexes live a quieter life, but human hunters are still a concern.

The ibex's droppings are cylinder-shaped and about 2 cm long.

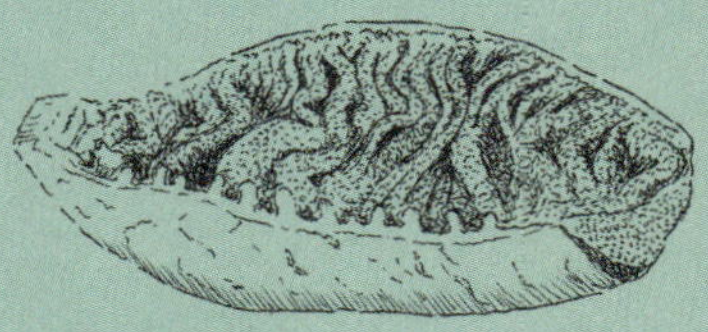

LIKE AN OPEN BOOK

The omasum is also known as the 'bible' because it's covered in thin, flap-like membranes, which make it look like a book with open pages.

HOOVES FOR HEIGHTS

The ability to reach steep areas means that ibexes can feed on shrubs and plants that other herbivores can't reach. The secret to these great climbers is their hooves, with hard sharp tips to grip every nook and cranny and non-slip pads to adapt to the terrain. They are the perfect hiking boots!

Hippopotamuses are the closest living relatives of whales and dolphins. This means they have more in common with whales than with other animals, such as pigs!

They graze for about four to five hours at a time. In that period, they can eat between 40 and 70 kg of grass.

HIPPOPOTAMUS
Hippopotamus amphibius

Animal: **VERTEBRATE** | Digestive system: **PSEUDO-RUMINANT** | Type of diet: **HERBIVORE**

Relaxation, not rumination

The stomach of a hippo has three chambers. In the first two (the parietal caecum sac and the forestomach), special bacteria break down the grass they eat. The third (the glandular stomach) is the 'true' stomach, which is where gastric acids are produced. However, unlike ruminants and other pseudo-ruminants including camels, hippopotamuses don't chew the cud. The food they swallow follows a one-way path through the digestive tract. This means that they get fewer nutrients, but they also need less energy!

FAVOURITE FOOD

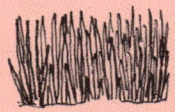
Grass

Water plants

Their diet is based on grass, which grows near the rivers and lakes where they live.

They also eat some water plants. Occasionally they have been seen eating carrion, even though their stomach is not adapted to digest meat.

WHEN DO THEY EAT?

They graze on dry land at sunset. They spend the day resting and keeping cool in the water or mud.

WHO EATS THEM?

Lions, hyenas and crocodiles attack young hippos but won't go anywhere near the adults. They are not just enormous and but also very aggressive! Their main enemies are human poachers.

1. oesophagus
2. stomach (with three chambers)
3. small intestine
4. large intestine
5. anus

These lips are for feeding, not flirting!

Hippopotamuses usually poo in the water. Male hippos spin their tails to spray the droppings around. It's their way of marking their territory.

OPEN WIDE!

A hippo's jaws can open very wide, up to an angle of 150 degrees. Although hippos are herbivores, they have enormous incisors and tusks that can be up to 50 cm long. They use them to eat greenery, defend themselves and fight with other hippopotamuses.

LAWNMOWER LIPS

A hippo's lips are just as important as its teeth when eating. They are enormous and very muscular and they are used to tear up the grass.

Unlike the dromedary or Arabian camel, which has only one hump, the Bactrian camel has two humps and originally comes from Asia.

68

A camel can drink up to 100 litres of water in only a few minutes.

BACTRIAN CAMEL
Camelus bactrianus

Animal: **VERTEBRATE** | Digestive system: **PSEUDO-RUMINANT** | Type of diet: **HERBIVORE**

Handy humps

The natural habitat of a Bactrian camel is harsh: rocky mountains and arid deserts with endless sand dunes. There are very few plants, limited water sources and extreme temperatures. Fortunately, this camel has two humps. These do not contain water, but are reserves of fat. The camel can use this fat for energy when conditions are tough.

FAVOURITE FOOD

 *Spiny plants*

 *Salty plants*

 *Dry grass*

 *Shrubs*

Anything goes!

Camels never waste a chance to eat plants wherever they find them. They don't mind feeding on the driest grasses or on prickly plants like cacti. They may also eat carrion meat, gnaw on bones or even nibble on ropes or fabric, if they have to.

WHEN DO THEY EAT?

At any time of the day. If they find food, they won't turn it down!

WHO EATS THEM?

Their natural predators are wolves and leopards, although those animals only hunt camels if they are very hungry. In reality, human beings are the main threat to wild camels.

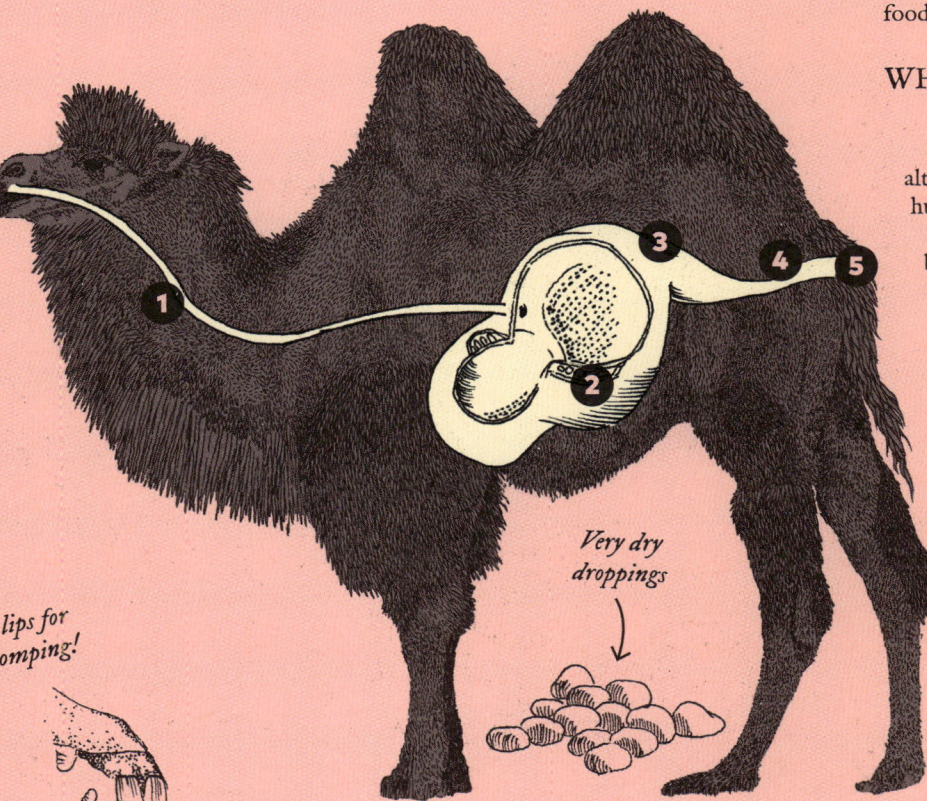

Very dry droppings

Double lips for extra chomping!

side view of teeth

1. oesophagus
2. stomach (with three chambers)
3. small intestine
4. large intestine
5. anus

FROM THE CAMEL'S MOUTH

The secret behind the camel's ability to eat any type of plant, no matter how prickly, is their extremely thick lips. The upper lip is also split into two halves that can move independently. These lips protect their strong teeth, which they can use as a weapon if necessary.

RUNNING ON EMPTY

The fat reserves in their humps allow camels to go for weeks without eating or drinking. Firm humps mean a full store of fat. But if the camel uses up these reserves and can't find food, the humps grow soft and floppy.

Hungry camel alert!

Llamas are pseudo-ruminants and belong to the Camelidae family, just like camels, although llamas don't have humps.

69

Llamas were selected and domesticated by Andean farmers from another member of the Camelidae family, called the guanaco, which, unlike the llama, lives in the wild.

LLAMA
Lama glama

Animal: **VERTEBRATE** | Digestive system: **PSEUDO-RUMINANT** | Type of diet: **HERBIVORE**

Spit and polish

A llama's saliva is full of enzymes that help to break down the plants they eat. But that's not all. It is also a defensive weapon! They are very territorial animals and when they feel their space is threatened, they will viciously spit as a warning. And sometimes it's not only saliva that they spit, but vomit too. Yuck!

FAVOURITE FOOD

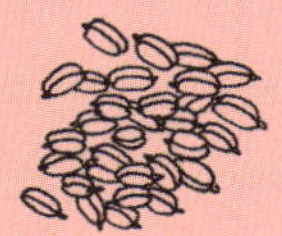

Roots · *Small shrubs* · *Grain* · *Lichen* · *Leafy plants*

WHEN DO THEY EAT?

They graze and ruminate throughout the day. Staying well fed in a tough environment like the Andes requires dedication!

Llamas live in the mountains of the Andes, at heights of 2,000 to 4,000 metres above sea level, and feed on the grasses and shrubs there.

Because they live in very dry places, they get a good share of the water they need from the plants they eat.

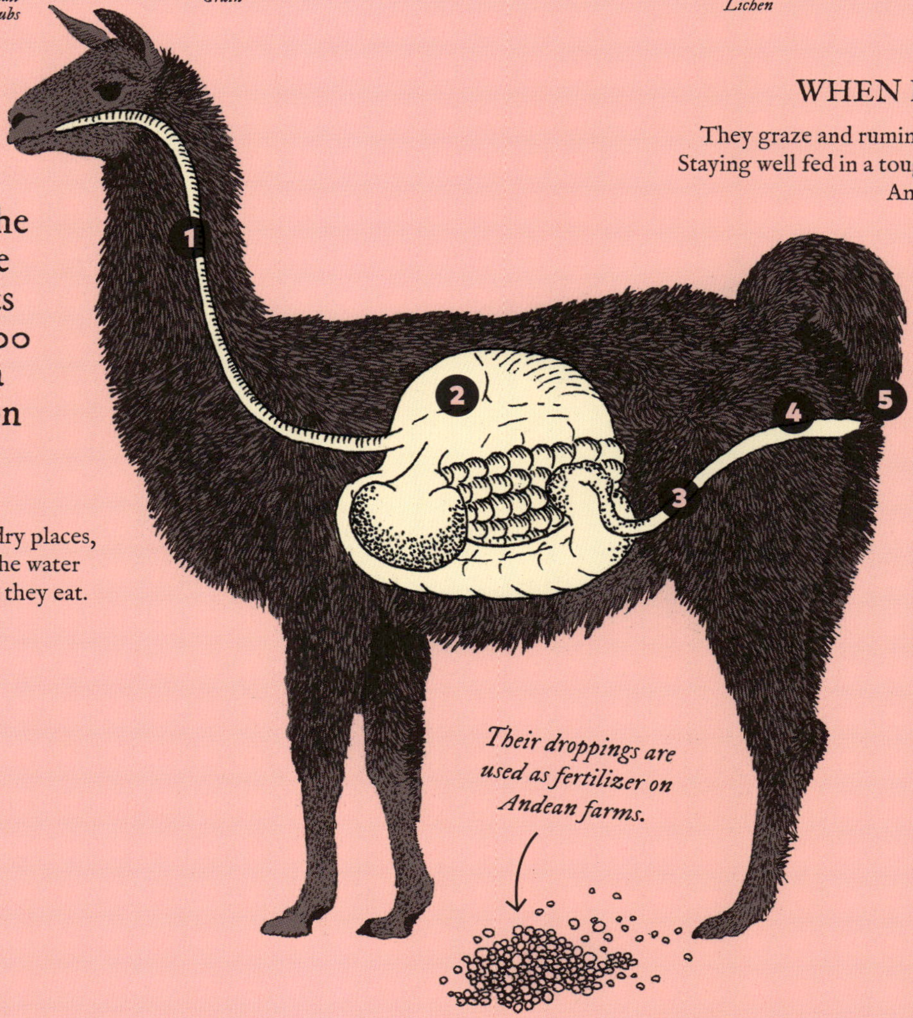

WHO EATS THEM?

Coyotes, pumas and ocelots. But llamas don't make it easy for them: they fight back with hooves and teeth (and spit)! Many Andean herders include llamas in their herds of sheep and goats to provide protection from predators.

1. oesophagus
2. stomach (with three chambers)
3. small intestine
4. large intestine
5. anus

Their droppings are used as fertilizer on Andean farms.

LLAMA LATRINE

One of the more unusual llama behaviours is their communal toilets or 'latrines'. Groups of llamas (usually including one male and several females) always poo in the same places, which are generally at a distance from their favourite grazing areas.

LLAMA TEETH

Although they are herbivores, llamas have well developed canine teeth. They don't use them for eating meat, but for fighting other llamas. Most of their fights are about territory or status within their social group.

These animals are the smallest members of the Camelidae family. They can reach 1.5 metres in height and weigh 40 to 50 kg.

70

The vicuña is the national animal of Peru. It even appears on the Peruvian national coat of arms.

VICUÑA
Vicugna vicugna

Animal: **VERTEBRATE** | Digestive system: **PSEUDO-RUMINANT** | Type of diet: **HERBIVORE**

Three stomachs in one

As pseudo-ruminants, the digestive system of a vicuña is different to that of true ruminants — its stomach has three chambers instead of four. The first chamber is used for fermenting grass, and is similar to the rumen of ruminants. The second chamber carries out jobs that are similar to the reticulum. Finally, the third chamber takes on the role of both the abomasum and omasum, but in pseudo-ruminants, it is shaped like a single tube and is not divided into two sections.

FAVOURITE FOOD

Shrubs *Grass*

Approximately two thirds of the vicuña's diet consists of pasture grass.
Unlike other members of the Camelidae family, they need water every day and usually live near rivers or lakes.

WHEN DO THEY EAT?
They graze by day in the lower areas of the region they live in.

WHO EATS THEM?
Their main predators are pumas, although they are also preyed on by foxes and dogs. They are farmed by humans for their wool, which is prized all over the world.

Like llamas, vicuñas poo in communal 'latrine' areas.

A mother's work is never done!

1. oesophagus
2. stomach (with three chambers)
3. small intestine
4. large intestine
5. anus

ABOVE AND BELOW

Vicuñas live at high altitudes, between 3,000 to 5,000 metres. Every day they go down the lower land where they can find their favourite grasses. As night falls, they return to higher ground. It is cooler there but their fleece protects them. They also avoid the places most visited by pumas, their main predators.

EATING FOR THREE

Vicuñas live in family groups made up of a male, a few females and their young. The females spend nearly all their time eating. A vicuña mother needs lots of energy as she is often pregnant while still nursing her previous fawn. So she needs plenty of nutrients, both for the baby inside her and for the young fawn's milk.

GLOSSARY

ANUS: the hole at the end of the digestive tract, where faeces (poo) comes out.

APEX PREDATOR: an animal that is at the top of the food chain, with no predators itself.

CARNIVORE: an animal that eats other animals.

CLOACA: the hole at the end of the digestive tract of a bird or monotreme, which is used for expelling faeces (poo), urine (pee), and for laying eggs.

COPROPHAGE: an animal that eats dung (poo).

CROP: a pouch-like organ near the throat, which is used by birds and some invertebrates to store undigested food.

DIGESTION: the process of breaking down food into nutrients so the body can use them.

DIGESTIVE TRACT: the long tube formed by all the organs in the digestive system.

DIURNAL: an animal that is most active during the day.

ENZYME: a type of protein molecule in the body that helps to break down food into nutrients.

EXCRETION: the process of getting ride of faeces (poo) and urine (pee) from the body.

FAECES: the waste (poo) that is left after digesting food, which must be pushed out of the body.

FORESTOMACH: another word for the proventriculus.

GIZZARD: also called the gastric mill, a part of the digestive tract used by birds and some reptiles to grind up food.

HAEMATOPHAGE: an animal that lives on the blood of other animals.

HERBIVORE: an animal that eats plants.

HOST: the animal that a parasite lives on.

INGESTION: the process of taking nutrients into the digestive system, usually by eating or drinking.

INVERTEBRATE: an animal without a spine (backbone).

MAMMAL: one of a large group of animals that feed on milk from their mothers.

MARSUPIAL: one of a group of mammals whose babies grow in a pouch outside their mother's body.

MONOGASTRIC: an animal with one stomach.

MONOTREME: one of a small group of mammals that lay eggs.

MYRMECOPHAGE: an animal that eats ants and termites.

NOCTURNAL: an animal that is most active at night.

NUTRIENT: a substance that a living thing needs to survive, grow and reproduce.

OESOPHAGUS: the tube that food must pass down in order to reach the stomach.

OMNIVORE: an animal that eats both plants and animals.

PARASITE: an animal that lives on the body of another animal and takes nutrients from it, often harming it at the same time.

PHARYNX: another word for throat.

PLANKTIVORE: an animal that lives on tiny sea organisms called plankton.

PREDATOR: an animal that survives by hunting other animals.

PREY: an animal that is hunted by other animals.

PROVENTRICULUS: also called the forestomach, a part of the digestive system that begins to break down food before it moves into the 'true' stomach.

PSEUDO-RUMINANT: an animal that chews the cud like a ruminant, but its stomach only has three chambers, not four.

REGURGITATION: the process of bringing digested food back up into the mouth.

RUMINANT: an animal whose stomach has four chambers, and which digests food by ruminating or 'chewing the cud'.

STOMACH: a bag-like organ of the body that is filled with gastric juices (acids) that break down food into smaller parts, so the body can use them.

SYMBIOSIS: a relationship between animals of two species, which harms neither of them and benefits either one or both of them.

VERTEBRATE: an animal with a spine (backbone).

Translated from the Spanish *Comepedia* by Lisa Agostini

First published in the United Kingdom in 2024 by
Thames & Hudson Ltd, 181A High Holborn, London WC1V 7QX

Original edition © 2024 Zahorí Books, Barcelona
Text © 2024 Víctor Sabaté
Illustrations © 2024 Aina Bestard
This edition © 2024 Thames & Hudson Ltd, London

All Rights Reserved. No part of this publication may be reproduced or transmitted in any form or by any means, electronic or mechanical, including photocopy, recording or any other information storage and retrieval system, without prior permission in writing from the publisher.

British Library Cataloguing-in-Publication Data
A catalogue record for this book is available from the British Library

ISBN 978-0-500-65386-9

Printed by GPS, Slovenia

Be the first to know about our new releases, exclusive content and author events by visiting
thamesandhudson.com
thamesandhudsonusa.com
thamesandhudson.com.au

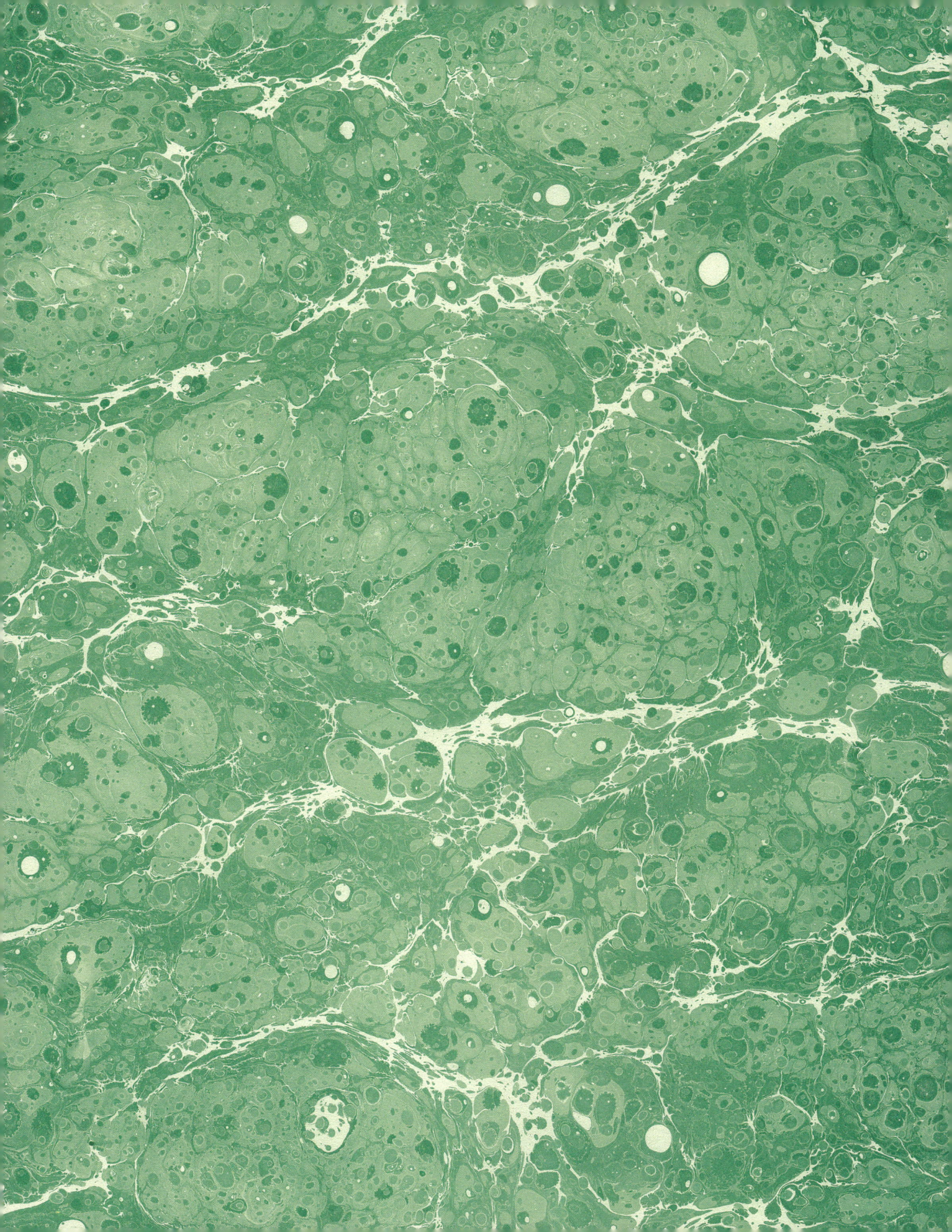

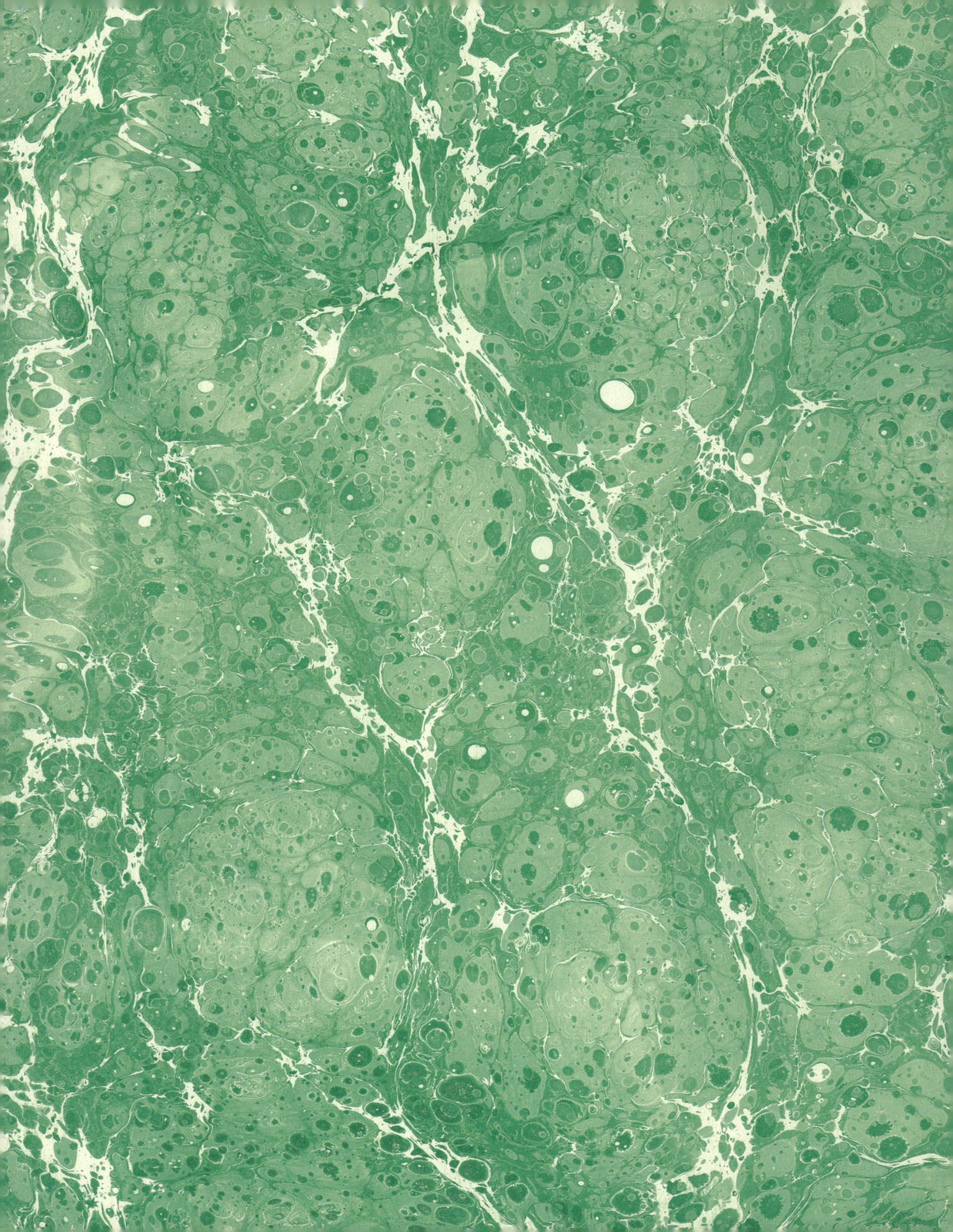